THE
REFERENCE
SHELF

PROSPECTS FOR ENERGY IN AMERICA

edited by ERIC F. OATMAN

THE REFERENCE SHELF
Volume 52 Number 3

THE H. W. WILSON COMPANY

New York   1980

THE REFERENCE SHELF

The books in this series contain reprints of articles, excerpts from books, and addresses on current issues and social trends in the United States and other countries. There are six separately bound numbers in each volume, all of which are generally published in the same calendar year. One number is a collection of recent speeches; each of the others is devoted to a single subject and gives background information and discussion from various points of view, concluding with a comprehensive bibliography. Books in the series may be purchased individually or on subscription.

Library of Congress Cataloging in Publication Data

Main entry under title:

Prospects for energy in America.

(The Reference shelf; v. 52, no. 3)
Bibliography: p.

SUMMARY: A compilation of articles discussing the energy problem and the pros and cons of possible solutions.

1. Power resources—United States. [1. Power resources] I. Oatman, Eric F. II. Series: Reference shelf; v. 52, no. 3.
TJ163.25.U6P76   333.79   80-15685
ISBN 0-8242-0646-0

PRINTED IN THE UNITED STATES OF AMERICA

CONTENTS

VII.   CONSERVATION: CHEAPEST FUEL OF ALL?

PREFACE

This collection of articles challenges the myth that the debate over America's energy future must appear either dull or incomprehensible to the average citizen. Because they helped create the myth, the maunderings of bureaucrats and the jargon of technologists are given little play here. Inevitably, some mumbo jumbo about megawatts, rad wastes, and the like remains in this compilation, but it survives in manageable patches, which readers can gloss over without fear of missing the major issues of the debate.

Getting an accurate idea of the energy picture requires one to cope with a host of new facts, surfacing almost hourly, that change both the picture and the public's perception of it. This aspect of constant change is worth a few paragraphs' examination, since it can make the energy debate seem as hard to follow as a shell game.

Try keeping track of crude oil prices, for example. Early in 1980, the price of a 42-gallon barrel of imported crude was quoted anywhere between $24 and $45, depending on whether the barrel was bought under contract or on the open market. To many, the new price was a shocker—it was more than 1,600 percent higher than the 1970 price of $1.30 a barrel. It is, however, widely conceded by experts in industry and government that such price hikes have predictable and, under the circumstances, beneficial effects on energy consumption: These increases encourage Americans to turn down their thermostats in the winter and cut back on unnecessary driving. Experts calculated that in 1980 conservation measures such as these would permit a drop in oil imports to the United States of at least 200,000 barrels a day. Without such gains from energy efficiency it was said that the United States would have to import 50 percent more oil than it does.

Right at this point the pricing shell game gets trickier. According to theory, lower demand for a product ought to keep that product's price from rising. But the nations that produce oil and sell it to the United States can keep oil scarce—and prices rising—by holding the supply of oil just short of demand. So, as 1980 began, the US Treasury Department had to predict that the nation's annual bill for imported oil would jump from $59 billion in 1979 to $81 billion in 1980—despite an estimated drop in imports of 73 million barrels!

The lesson? It is that the energy picture is not a shell game and that even its most baffling details make sense when looked at closely. The endlessly rising price of crude oil reminds us that the nation's reliance on a dwindling resource—nearly half of which it must import—has become a costly habit, which Americans pay for with almost every purchase. The prices of gasoline (up 35 percent in 1979) and home heating oil (up 36 percent in 1979) are some obvious manifestations of the nation's dependence on foreign oil. Less obvious, perhaps, is the tie-in with the price of food (up 11 percent in 1979), but the production and distribution of food devour huge amounts of energy.

The skyrocketing cost of energy has put Americans in a squeeze, leaving most of them with less money to spend on nonessential items. From an energy standpoint, economizing through conservation can be desirable, as this volume makes clear; yet, enough corner-cutting at the marketplace is bound to increase the ranks of the unemployed. For instance, reduced new-car sales in 1979 resulted in substantial layoffs in the automotive industry. And in December 1979, when foreign oil producers raised the minimum price of a barrel of oil from $18 to $24, government economists predicted 250,000 Americans would lose their jobs in 1980 as a result of this hike.

Higher oil prices, however, are neither the only culprits in the nation's energy picture nor the sole cause of the nation's extraordinary inflation rate. Economists blamed only a quarter of the nation's 1979 inflation rate of more than 13 percent on higher energy prices, although the price of foreign oil more than doubled that year. Experts point out that the nation owes its misfortunes with energy not just to a shortage of cheap oil but to a host of causes, including disarray in the nuclear power industry, environmental roadblocks to coal exploitation, and the inability or unwillingness of Americans to use energy more efficiently. Then, too, as several articles in this volume suggest, government policies (and, in some areas, their absence) appear to be as much a part of the nation's energy ills as they are of its cure.

The volume is divided into seven sections. The first two, "Stalking the Problem" and "Inventing Solutions," carve out a definition of the energy crisis and examine several competing suggestions for dealing with it.

With the third section, "The Nuclear Dilemma," we leave the general debate over policy behind and begin to evaluate specific alternatives to our dependence on increasingly expensive oil. Section IV, "Coal and Natural Gas," examines two energy resources whose increased use is as problematic as nuclear energy. Section V, "Alternative Fuels," explores the feasibility of "synfuels" and combustible alcohol distilled from plants.

Experts remain split on the role that solar power will play—or should play—in the nation's future "energy mix," as several articles in Section VI demonstrate. No "menu" of possible responses to the challenge of an oil-short world can avoid taking into account the solar energy locked in biomass and wind, in rolling rivers, and in ocean warmth. This section takes a look, too, at the energy of the tides and of the super-heated water deep in the earth's crust.

Conservation, however, may be the cheapest fuel of all. As the 1980s began, most major studies—including those of Harvard's Energy Project and of the National Academy of Sciences—were agreed on that point. The book's final section explores conservation's promise and reviews some of the benefits that at least one town is already reaping from it.

The compiler wishes to thank those authors and publishers who were kind enough to let their materials be reprinted in this book. Once more, the Forbearance Award goes to his wife, Jane, and to his daughter, Alison, for managing the crisis this energy project brought home.

ERIC F. OATMAN

April 1980

I. STALKING THE PROBLEM

EDITOR'S INTRODUCTION

In the often bitter debate over America's energy future, all sides agree that the nation faces a momentous challenge. However, the suggestion that the combatants define that challenge is likely to close off further opportunities for consensus. Americans were not shocked into an awareness that a challenge existed until the Arab oil embargo of 1973. So, perhaps at this stage such fractiousness is to be expected, even welcomed. In any case, the workings of the nation's energy supply systems are difficult for the public to grasp. Their interrelationships are further complicated by the pressures of the demands for energy—distorted, as they are, by government regulations on price and distribution. Consequently, if the participants in the energy debate seem to be talking past each other, they often are. Their backgrounds may determine the way they view the mechanism of energy supply and demand, and the data they use to explain it may be the data they are most comfortable with.

Nonetheless, a clear picture of the challenge of an oil-short world has emerged, most comprehensively out of a six-year study at the Harvard Business School. (An excerpt from that study's final report, *Energy Future* [Random House, 1979], closes Section II.) Daniel Yergin, an historian and a co-director of the Harvard group, gave readers of the New York *Times Magazine* a glimpse of that study's findings in 1978. His analysis of the energy crisis, which many considered alarmist before the 1979 Iranian revolution set the stage for a second "energy shock," is today viewed by most observers as balanced and fair. Yergin's analysis begins this section.

In the second selection, "The Nation's Energy Chain," New York *Times* energy reporter Anthony J. Parisi describes the interrelationships of the nation's energy supply systems. The final selection, by James Nathan Miller, a roving editor of *Reader's Digest*, suggests one reason why the energy crisis is so hard even for experts to comprehend. "A dozen or so private companies," Miller charges, "exercise almost total control over what facts the country is allowed to see in determining its energy future."

WHAT THE ENERGY CRUNCH MEANS[1]

It need not happen. But, unless we bestir ourselves, it most likely will. By 1985 or 1986 or 1987, if present trends continue, we'll be staring at an energy crisis far worse than the one we went through in the early seventies. And then our present boredom with the energy problem, and with the Carter Administration's efforts to cope with it, will seem like complacent sleep.

In that event, the reality we wake up to is apt to be frightening. Prices will double or triple, in real terms, within a short time. The standard of living of every American will nosedive. The international monetary system will shudder and shake. Industrial nations will be pitted against each other in a bruising scramble for oil. The Western alliance could be shattered. In a number of countries, democracy itself might not be able to survive.

There is one sure way to forestall this disaster, and that is for Americans to get their oil thirst under reasonable control. The United States dominates the world oil market. We use a third of all the oil used in the world every day. American cars and trucks alone use a seventh of all the oil used in the world every day. Our ... trade deficit, and the consequent weakening of the dollar, is mainly the result of our growing oil im-

[1] Excerpts from the article "The Real Meaning of The Energy Crunch," by Daniel Yergin, member of the Energy Research Project at the Harvard Business School, coauthor of *Energy Future* and author of *Shattered Peace*. New York *Times Magazine*. p 32+. Je. 4, '78. Copyright © 1978 by Daniel Yergin. Reprinted by permission of The Helen Brann Agency, Inc.

ports. Yet in the present atmosphere of careless optimism about our oil supplies—"We're doing all right," people say— the goal of sensible restraints on our consumption appears harder and harder to attain.

Let us give the optimists their due. They have their reasons. The world does continue to function. . . . Some recent projections and studies have been widely played in the press under such headlines as "Experts Dispute Administration, Doubt Energy Crisis in the 80's." People even say there is an oil "glut." . . .

But the famous "glut" amounts [in 1978] to only 2 million to 3 million barrels a day, less than 5 percent of world production—nothing more than a temporary slack. And [in December 1977] at Caracas, the Saudi oil minister, Sheik Ahmed Zaki Yamani, qualified the no-hike-now decision with the statement: "Once the [present] surplus is eliminated, neither the United States nor any other superpower will be able to bring about a freeze in oil prices." The real price of oil, he said, will almost certainly double within a decade.

Yamani was being realistic. He was pointing to what is clearly in prospect for the late 1980s—the energy crunch, sometimes called the gap.

This coming gap has been the subject of heated discussion among energy specialists. . . . There are two ways of construing it. One is the notion of an absolute breach between supply and demand—not enough oil left in the ground to meet the world's needs. The Administration has used this dramatic imagery to try to sell its energy program. But it is misleading. It is very hard to demonstrate that the world will run out of oil in the next decade or two. World production grew by 70 percent between 1966 and 1974; in the same period, world reserves grew by 80 percent. What should concern us, although it has been neglected in most public discussions up to now, is the second meaning of the gap.

Under this definition, the gap will occur in the middle or late 1980s. Production of non-OPEC [Organization of Petroleum Exporting Countries] oil will, by then, be running at full speed. And production of OPEC oil will be approaching certain limits. These may be physical limits—the capacity of the wells, pipelines, docks and other equipment in place. Or they may be limits imposed by OPEC governments for political or economic reasons. In either case, worldwide demand for oil will begin to run up against the limits of worldwide production, and we may begin to see interruptions of supply.

With or without such interruptions, we can be sure of something else: With supplies that tight, it will again be a seller's market. Prices will go up, rapidly and dramatically.

Here, then, is the real meaning of the energy gap—a point at which we would witness a very sharp breach between demand and supply *at reasonable prices.* We would see a sudden shooting up of prices from those the world can bear to those it can't. . . .

We are talking about a predicament that, to be avoided, must be grappled with now. And, contrary to much current publicity, it should be seen not as a new crisis but as the sharpening of a situation that emerged in the late 1960s, made its impact felt in the years 1970–74, and is likely to continue into the 1980s and 1990s.

Up until the late 1960s, the network of oil production, transportation, refining and distribution was under the control of the major oil companies (the Seven Sisters) and an assortment of independent oil firms. The oil was delivered on time and cheaply, and the consuming nations were more or less happy to let the companies handle the whole thing. As for the producing countries, they were not in a strong enough position to do much more than hand out concessions and collect royalties. But in the late 1960s, the international energy system was transformed.

The industrial world was by then in the midst of a rapid shift away from coal to very high dependence on oil. Who wanted to stay with coal? It was bulky, it was dirty, and it represented old forms of social conflict that most everybody wanted to forget. How much easier to turn a valve for oil than to shovel coal! Between 1955 and 1972, the share of Western Europe's energy needs met by coal declined from three quarters to little more than a fifth, while the share met by oil increased from a fifth to three fifths. A similar change, starting even earlier, occurred in the United States.

The impact of this change was magnified by another factor. The industrial nations consumed more and more. In 1976, North America, Western Europe and Japan used 65 percent of all the oil consumed in the world. But they increasingly used other people's oil, for they had only one tenth of the world's reserves. (Meanwhile, the Middle East and Africa, with two thirds of the world's known reserves, consumed only 4 percent of the oil.) Few people realize how recent this change has been. The United States was still the world's No. 1 producer in 1964. The next year, it was overtaken by the Middle East. Before the decade was out, the Middle East was producing a third more oil than the United States.

At the same time, the West was losing its "petroleum insurance policy." This had taken the form of oil wells in place but not in production—what the industry calls spare capacity. So long as such capacity existed in the United States, a squeeze play in the Middle East could not be very effective, since the unused wells in the United States could be put back into production. This is what happened during the Arab-Israeli war of 1967, thus aborting an attempted Arab oil embargo.

But the United States, at that time, was already running out of spare capacity. American production peaked in 1970 and began declining thereafter, and spare capacity disappeared. The import quotas that had restricted the amount of foreign oil in the United States were scrapped by Washington, and the hungry American market became part of the world market. Hence the confidence of Sheik Yamani's remark to an Exxon executive in 1971: "You know the supply situation better than I do. You know you cannot take a [Middle East] shutdown." The insurance policy was void.

The OPEC countries did not shy from capitalizing on their advantage. They were no longer content merely to collect rents. They wanted the bulk of the income and they wanted control, and between 1969 and 1973, ever more sophisticated and assertive, they took both. The international oil companies no longer ran the show; they were reduced to little more than ticket takers. They have done rather well, even so, but control was now in the producers' hands. And once they had knocked over the companies, the producers began to exert power directly over the consuming countries.

Most of the industrial countries suddenly found urgent reasons to overhaul their Middle East policies. They were suddenly more than willing to pay court to the new bankers of the world economy. Those bankers also cast themselves as the champions of "third-worldism," using their oil weapon to demand a redistribution of power and product in the world. (That poor countries without oil have been hurt worse than anyone else by OPEC's price policies is something OPEC prefers to overlook.)

The transformation might not have been so thorough had it not been for one last factor. OPEC is composed of thirteen nations with competing interests and no lack of mutual suspicion. If their oil reserves had been distributed equally among them, they might have found themselves constantly at odds. Fortunately, from their point of view, geology decided otherwise. Saudi Arabia, a country with one quarter of the world's oil reserves, and with natural conditions permitting extraordinarily cheap and easy exploitation, has emerged as the cartel's dominant partner and leader. . . .

It took a series of events, including the [October 1973] Yom Kippur War, the [ensuing] Arab embargo and Western panic-buying, for these changes to become clear to all concerned. Now that the producers know how considerably power has shifted to their corner, they will not need new Arab-Israeli hostilities to respond with sharp price hikes or interruptions of supply in pursuit of what they perceive to be their own interests.

This crisis-prone relationship is likely to persist into the 1980s. In the past year or so, there have been a number of projections for the oil balance in the . . . [1980s]—by the Central Intelligence Agency, Exxon, Shell, the Workshop on Alternative Energy Strategies at the Massachusetts Institute of Technology, the International Energy Agency and various arms of Congress. A newspaper reader might be left with an impression of sharp disagreement. On the contrary, there is a surprising degree of consensus. The controversy has focused on just one aspect of the CIA's studies—the prediction that the Soviet Union, a not insignificant oil exporter, will become a major oil importer because of its domestic production difficulties and will thus make the oil market even tighter.

The evidence both for and against the CIA's case remains slim. The Russians themselves probably don't have a very clear idea of where they will be in the 1980s. This question aside, the CIA is right in step with the other projections. A consensus view would be something like this:

World demand for OPEC oil is now between 28 and 31 million barrels a day. Available OPEC capacity is 33 million barrels a day. If pushed very hard and willing to respond, the OPEC nations could go as high as 38 or even 39 million barrels a day. . . .

[By 1988] however, the demand for OPEC oil will be at least 42 million to 45 million barrels a day. Meanwhile, OPEC's physical capacity, if pushed to the utmost, might have grown to around 45 million barrels a day. That will make the supply situation extraordinarily tight.

Moreover, it is not clear that the OPEC nations could or would want to produce at the outside limit of physical capacity. Some OPEC spokesmen have been forthright in warning against such an assumption. "It is felt necessary that the life span of OPEC's hydrocarbons reserves should be extended as long as possible," OPEC secretary general Ali M. Jaidah said . . . [on May 8, 1978], predicting a production level for the mid- and late 1980s of only 42 million barrels a day, if not lower. And he made clear that even the 42 million level could be expected only if made "economically justifiable"—that is, in response to much higher prices.

What will follow from this crunch? The period 1970–73 gives us some idea of what to expect. The tightening of demand and lack of spare capacity would once again strengthen the hand of the price hawks . . . [in OPEC]. By this time, the OPEC governments would have extended their reach not only into the transport, refining and marketing of oil but into the international trade in natural gas. They might not act as overtly as the last time but rely on "impersonal" market forces to push prices up.

The effect of climbing prices would be grave. There

would be a resurgence of hyperinflation. At the same time, there would be another major reduction of investment and purchasing power in the industrial world, with catastrophic impact on growth and employment rates, as well as on that basic confidence in the system on which our type of economy depends. The strains on the international payments structure would be intense; countries like Italy, France and Japan might well experience intolerable balance-of-payment crises. The developing countries would feel those strains most acutely, with devastating effect on their development plans. In such circumstances, one would hardly expect the private banking system to weather the storm with any great degree of comfort. In sum, the economic consequences would likely be a major recession, even a world depression.

The political consequences would be no less serious. It is economic growth in the years since World War II that provided the solution to the conflicts and internal crises that liberal democracies experienced in the 1920s and 1930s. How long can such systems survive without the solvent effects of growth? High unemployment, hyperinflation, the struggle of various groups to maintain their relative and absolute positions—all this would make it extremely difficult for social conflicts to be mediated through the traditional play of democratic politics.

Conflicts would arise not only within the Western nations but among them. All other aspects of their relationship would be overshadowed by a bruising scramble for oil. As lights flickered, prices shot up and publics panicked, political leaders would find themselves unable to respond to anything other than immediate political imperatives. That some nations would be better off than others—the United States less vulnerable than Western Europe and Japan, West Germany in a better position than Italy—would only make matters worse.

How long could the Western security and trading network survive such pressures? Certainly, the OPEC nations would be able to play various countries off against each other. Also, the potential Soviet role cannot be ignored. If it succeeds in solving its own oil production problems, the Soviet

Union might well seek to use its oil exports to gain political leverage over West Germany and Italy. Conversely, if the CIA's prediction of a Soviet oil shortage turns out to be right, the Russians might well be desperate enough to take bold risks in trying to extend their influence over oil producers in the Persian Gulf area. Without being alarmist, it must be recognized that wars have been started for less reason than that.

In the late 1940s and 1950s, abundant, low-priced oil helped Western Europe and Japan make their spectacular recoveries within the Western system. In the latter half of the 1980s, expensive oil could bring that system crashing down. We do not have to go by forecasts alone. A precursor of what to expect is what happened during the energy crisis of 1973–74, when the balance of supply and demand was still better than the one we are heading into.

In the first few months after the 1973 oil embargo, the very basis of the European Community was threatened by the refusal of some members to share oil with the Dutch, who had been singled out by the Arabs for a complete embargo. (This situation was remedied when the Dutch reminded the French that a goodly share of France's natural gas happened to come from the Netherlands.) At the same time, the basis of NATO was shaken when frightened European governments denied landing rights to American planes resupplying Israel.

These conflicts have since been muted, as the industrial nations have sought to work together through such forums as the International Energy Agency. But the present calm may only be temporary. One need look no further than the dispute stirred by President Carter's nuclear energy policy to realize the shakiness of the accord on oil.

The United States is fighting a rather late battle against the development of certain nuclear technologies—the recycling of nuclear waste and the breeder reactor—that increase the likelihood of the proliferation of nuclear weapons. This has pitted the United States against other industrialized countries, including France, West Germany and Japan, that have been developing these technologies for their own use and for export. Fundamentally, these countries believe the new American policy would deny them the tools they require to reduce their dependence on OPEC oil. This, in their eyes, poses a threat to their very survival. (For that reason, no doubt, they are less sensitive to the dangers of nuclear proliferation.)

The very preponderance of the American role in the energy field engenders suspicions among our allies. In some quarters, the American antinuclear policy is regarded as an expression of commercial rivalry. There are still some who wonder whether the United States manipulated the 1973 crisis to undercut the economic competitiveness of Europe and Japan. Many wonder whether our special relationship with Saudi Arabia represents an effort to preempt the Saudi output for the United States when the crunch arrives. The fact is that Washington's strong stands on international energy and nuclear matters lack credibility as long as the United States remains without a domestic energy policy. For, to the other countries, this means that the United States continues to gobble up the world's oil while asking them to hold back on their efforts to reduce their own far greater vulnerability to the hazards ahead.

There is another danger, this one so overwhelming that officials prefer to pretend it doesn't exist. . . . If the increased demand for OPEC oil projected for the 1980s is to be met, it can be done only by raising Saudi production to 16 million or even 20 million barrels a day.

The Saudi oil industry today has a physical capacity of up to 10.5 million barrels. By 1982, it will have a capacity of up to 14 million barrels. How much higher it can go is not clear. Whether it can be stretched . . . to meet the demand placed on it will have enormous international impact.

Dependence on Saudi Arabia to such a degree would make that country not only the linchpin of the OPEC cartel but the linchpin of the world economy—a surprising responsibility for a nation of perhaps five million, undergoing what may be the most rapid and total modernization in world history. The industrialized countries would be placed in a very precarious position. Their well-being would be affected not only by decisions in Riyadh but by a host of imponderables.

There could be a natural disaster. Or an accident or a ter-
rorist strike in the oil fields or in the narrow straits leading to
the Persian Gulf. Or the growth of the Soviet presence in the
region, some new twist of the Arab-Israeli conflict, or a struggle for preeminence among Saudi Arabia, Iraq and Iran. Or a
shift in the outlook of the elite that runs the Saudi kingdom.
Or—and this is the very real danger that Western officials
prefer to close their eyes to—a coup in Riyadh and the ac-
cession to power of a radical like Libya's Colonel Qaddafi.

The present Saudi leaders deserve their due. Except for
their conduct during the 1973-74 crisis, they have been
rather cautious about brandishing the oil weapon. Granted,
they presumably recognize that it might not be in their interests to threaten economic warfare against nations that more
or less underwrite their security. A Qaddafi in the Arabian
peninsula, on the other hand, is not likely to be that con-
strained. The Libyan dictator has been prone to frequent and
arbitrary cutbacks in production and changes in the rules, for
a whole range of political, ideological and emotional reasons.
A Saudi Qaddafi might well follow the same irrational pattern, particularly since the Saudi regime earns so much more
from its oil than it can possibly spend at home. The coming to
power of such leadership in Saudi Arabia would create tremendous uncertainty around the world.

And this worry is not the only one. A similar upheaval in
Iran, the No. 2 OPEC producer, or some other key country in
the region, would also have very serious consequences. [And
it did, in 1979, when Iranians ousted the Shah and Iranian oil
production dropped. On November 4, Iranians took over the
US embassy in Tehran, holding Americans hostage, and the
United States countered by refusing to buy any oil from
Iran.—Ed.]

Geology need not be destiny. . . . The odds could be diminished by what occurs on the supply side.

But there does not seem to be much reason today to be
optimistic about these alternatives. Alaskan production in a
decade or so will only make up for declining production in

the Lower 48 States. The North Sea oil fields could reach 5 million to 6 million barrels a day by 1985, but, barring some major new finds, their output will then begin to go down. The only major new oil strike of the 1970s was the one in Mexico, and production there is being held back by political and technical problems.

Nuclear energy? Projections for its contribution have been consistently lowered since 1973. A tangle of barriers stands in the way—cost, technical problems, environmental risks, doubts about safety, and, most recently, the dispute over nuclear proliferation. To put it simply, nuclear power is in the grip of a paralyzing stalemate. [See Section III, below.]

Coal? It has become increasingly clear that there are plenty of obstacles to a return to primary reliance on that resource. [See Section IV, below.] Solar energy? It could probably make a much greater contribution than conventionally thought, but many more incentives are required. [See Section VI, below.] In sum, prospects for avoiding the crunch through non-OPEC oil and through alternative supplies do not appear to be good.

But what about changes in world demand? If the projected need for OPEC oil in the 1980s is brought down to below the critical figure of 42 million barrels a day, the whole picture is altered.

The demand for oil would drop if economic growth rates proved to be even lower than the already lowered recent forecasts. But that is not exactly a happy way to solve the energy problem. It would mean high unemployment and other economic and political consequences of the same kind that would be produced by an energy gap. . . .

Unhappily, complex and fragmented problems that stretch far ahead and involve foreign countries tend not to get the sustained attention they require. As Immo Stabreit of the International Energy Agency has pointed out, "Energy policies are politically risky, because the danger with which we are dealing is 8, 10 or 12 years away, whereas our governments face re-election every four years." In other words, the glory and reward for successful avoidance of disaster, or for providing a better way of life, cannot be attained during the politically relevant period—before the next election. "There is, therefore, a strong built-in urge to dodge the issue."

Dodge it now, and the United States and the other Western countries will pay later. That is the challenge before the West—and primarily before the United States, with its overwhelming impact on the world oil market. For here are posed, as one peers into the 1980s, issues of survival that, in their own way, are no less fundamental than those of the nuclear arms balance.

THE NATION'S ENERGY CHAIN[2]

The immediate energy problem . . . revolves around the nation's troublesome dependency on oil imports, which started to become a significant factor in the United States only about ten years ago.

Since then, Americans have not only seen the price of foreign oil multiply tenfold but have also suffered two extended periods of supply cutoffs from abroad: the current one [1979] and the five-month Arab oil embargo that spanned fall and winter 1973–74. But the nation's energy woes inevitably entail far more than oil imports. . . .

Like the separate wheels and cogs in a complex train of gears, the energy system is an intricate and interlocking network of power sources. Anything that affects one source ultimately affects the rest—or, put another way, when something troubles one aspect, adjustments are generally required elsewhere in the system.

The following is a description of that energy system, its components and their potentials and limitations. The figures were assembled from an assortment of government and industry clearing houses.

[2] Excerpts from the article "Nation's Energy Chain: Fuel Sources Intertwine," by Anthony J. Parisi, staff reporter on energy and related subjects. New York *Times*. p A 12. Jl. 16, '79. © 1979 by The New York Times Company. Reprinted by permission.

Total Energy Use

The United States consumed an estimated 77.6 quadrillion, or million billion, British thermal units of energy . . . [in 1978], or almost one quarter more than Western Europe and Japan combined. One BTU is the amount of energy needed to raise one pound of water by one degree Fahrenheit, and energy specialists usually call a quadrillion of those measures a "quad."

Happily, 365 million barrels of oil contain roughly two quads of energy; that simplifies the conversion of quads to millions of barrels of oil a day, another common large energy measure. In 1973, the year the Arab embargo began to point up this country's energy vulnerability, Americans consumed 74.6 million quads.

Thus, the overall growth in consumption has been quite meager—just 4 percent over five years. However, domestic energy production has fallen in this period, and so imports have had to account not only for the growth in consumption but also for the dropoff in domestic supplies.

Oil, the Biggest Import

Oil satisfies the lion's share of the country's energy needs and is by far the single biggest energy import. The nation consumed about 37.8 quads of oil in 1978, or 18.7 million barrels a day. That satisfied more than 48 percent of the country's overall energy needs. Of the total, 8.1 million barrels a day was imported.

In 1973, Americans consumed 17.3 million barrels a day, of which only 6.3 million was imported.

Domestic output of oil and gas liquids, byproducts of natural gas production, slipped from 11 million barrels a day in 1973 to 10.6 million . . . [in 1978], even though the oil industry drilled almost twice as many wells in 1978 as it did in 1973.

The reason for the decline, energy experts say, is that the United States is running out of oil. Accelerated drilling seems to have slowed the decline somewhat. But at the present rate of consumption the nation's proved oil reserves amount to only a four-year supply, though it would be physically impossible to extract the oil that quickly.

New reserves are found each year, of course, but the last time the industry discovered more oil than it produced was in 1966. Reserves peaked in 1970 and have been declining ever since.

Nonetheless, President Carter hopes to slow still further the decline of production by phasing out price controls at the well. But with reserves dropping steadily, few oil analysts now expect the decline to be reversed. Most say that, at best, it may level off for a while before resuming its downward path.

Natural Gas

Gas is the second most prevalent energy source in the United States. The 20.3 million million cubic feet that Americans consumed . . . [in 1978] provided 19.8 quads of energy, or about 25 percent of the total. That is the equivalent of close to 10 million barrels of oil a day.

About 5 percent was imported, the vast majority from Canada by pipeline; the rest came from abroad, in liquefied form.

Gas imports were about the same in 1973, when consumption was at 22.5 quads: In other words, domestic gas production has also slipped, and like oil, reserves have slipped even faster.

Except for the inclusion of the reserves on the North Slope of Alaska, which cannot be tapped until and unless a gas pipeline is built to carry them south, reserves have fallen steadily since the peak year of 1967. They now stand at 203 million million cubic feet, which is a ten-year supply at the current rate of consumption.

However, though they are pessimistic about future oil production, many energy experts are optimistic about gas. [For further information, see Eugene Luntey's article

"What's Wrong With Gas," below in Section IV.] They think the industry may find a lot of deep deposits once considered too costly to probe. Their hope rests largely on [. . . 1978's] natural gas legislation, which sharply raised the price for such discoveries and opened the way for the eventual decontrol of all gas.

Coal, the Most Abundant

Although it is by far the most abundant indigenous energy source, coal ranks only third in consumption. . . . [In 1978], it provided Americans with an estimated 14.1 quads of energy, or the equivalent of roughly 7 million barrels of oil a day. That satisfied barely 18 percent of total energy needs. In 1973, coal provided 13.3 quads.

These days, the United States exports coal. . . . [In 1978], even though production was crippled for three months by the miners' strike, the industry exported about 6 percent of the 653.8 million tons of coal that it mined, along with an additional five million tons of anthracite. . . . [In 1979], it expects to produce 724 million tons and export about 50 million.

At the present rate of consumption, the nation's known coal reserves would last at least three hundred years and a lot more remains to be found. But environmentalists' objections, some expressed about mining but most about where the coal is burned, are restraining the growth in consumption. Coal can be converted into oil and natural gas, but the costs are still considered prohibitive and the conversion process itself is fraught with environmental problems.

At one time, Washington was talking about doubling or even tripling coal production by 1985, but now most energy experts expect coal consumption to increase no faster than its current, rather modest rate. [See articles by John I. Mattill and Bob Tippee in Section IV, below.]

Renewable Sources

Solar hot-water heaters and space heaters, rapidly gaining in popularity, still do not provide a measurable amount of energy in the total picture, and neither do such other renewable forms of energy as firewood or grain alcohol, which can be mixed with gasoline to make "gasohol." [For further information on gasohol and synfuels, see Section V, below.]

However, hydroelectric power plants, another renewable form of energy that is based on yet another renewable form of energy, supplied Americans with three quads of energy in 1978. That is the equivalent of about 1.5 million barrels of oil a day, or almost 3.9 percent of the total.

Energy analysts say that most of the nation's water-power potential has been tapped, but many expect other forms of renewable energy to start rapid growth. They hope solar, hydropower, geothermal, firewood and special "energy crops," conversion of refuse into energy and all the rest will account for 20 percent of the nation's energy requirements within twenty years. [For further information on renewable energy sources and the viability of their use, see Section VI, below.]

Proponents of renewable energy complain that present tax laws and other institutional factors inherited from the days of plentiful energy give nonrenewable sources an advantage over renewable ones. They therefore argue that the Government should subsidize renewable forms of energy to offset the institutional barriers.

In a book called *Energy Future* ... [Random House, 1979], even the Harvard Business School has embraced this once-radical position. The authors concede, however, that the potential for renewable energy sources remains unclear and, as a result, they argue that Government measures to encourage greater energy efficiency—more insulation, higher auto gas mileage, better appliances and so on—would be even more rewarding. [See Section VII, "Conservation: Cheapest Fuel of All," below.]

Nuclear Power

Atomic power supplied 2.9 quads of energy in 1978, or more than 3.7 percent of the total. Nuclear power plants now account for about 9.5 percent of the nation's total installed generating capacity. But because they operate day in, day out, regardless of fluctuations in the demand for power, they furnished almost 13 percent of the kilowatt-hours consumed in the United States . . . [in 1978].

Despite the accident at Three Mile Island in Pennsylvania, nuclear power continues to grow at an annual rate of about 11 percent, as plants ordered years ago keep coming on line. The industry expects at least half again as much nuclear power to be available by the end of 1988 as there was at the end of 1978.

Beyond that date, the picture gets cloudy. Plants under construction will presumably be completed, but new orders for plants slowed to a trickle five years ago and, unless they pick up again soon, the growth in nuclear power will halt before the end of the 1980s. Environmental, safety and even economic objections will have to be overcome first. [See Section III, "The Nuclear Dilemma," below.]

Synthetic Fuels

In addition to coal, the United States has vast deposits of oil shale that can be converted into petroleum. Both costs and environmental concerns are holding this effort back. Although there have been numerous proposals and experiments, commercial conversion of coal and shale oil into synthetic fuels has not begun. Many persons familiar with the matter consider the problems so overwhelming that they believe the conversion can never begin without Government help.

The President has proposed such a program [see "Phase 2 of Carter's Energy Plan," in Section II, below.], but even with a big Government push, synthetic fuels are not expected to provide more than 1 or 2 million barrels a day by 1990, or less than 5 percent of the nation's current energy requirements.

Nowadays, the term *synthetic fuels* has come to encompass solid, liquid and gaseous fuels made from renewable energy sources as well. Alcohol distilled from grain is one currently popular example. Others include methane made from natural wastes, fuel pellets fashioned from refuse and hydrogen dissociated from water.

As with nonrenewable synthetics, most of the renewable forms also face tough cost hurdles. Generally, though, they pose far fewer environmental objections. [For further discussion of synthetic fuels, see Section V, below.]

WHO HAS THE FACTS?[3]

Last year [1979], the State of Washington, conducting an antitrust suit against ten of the country's largest oil companies, assigned a researcher, Jack Andresen, to look through Department of Energy [DOE] files for data on the companies' imports. [The suit alleges that the companies used the 1973–74 Arab oil embargo as an excuse to raise gasoline prices.]

Andresen says:

Two DOE officials were assigned to help me. Both were former employes of Exxon, a defendant in the suit. The key set of documents was missing from the files; they had no idea why. They answered every substantive question about the missing documents with, "Sorry, I don't know." When I asked to see their boss, he refused to talk to me. The State of Washington sent a second official East to see if he could get coöperation from another office within DOE. He got none.

The US public faces the same problem: how to pierce DOE's secrecy screen to get the facts on energy matters. To understand the seriousness of the problem, consider two past Congressional decisions.

□ In 1973, Congress was debating how to transport Alaska's newly discovered oil to the Lower 48 States. The oil companies wanted to haul it by pipeline across Alaska to the Pacific, and from there by tanker to the West Coast. But an analysis by a Washington economic study group raised doubts about the companies' motives.

The West Coast, said the analysis, was already well sup-

---

[3] Article entitled "Let's Get the Facts," by James Nathan Miller, roving editor. *Reader's Digest.* 116:90–2. Ja. '80. Reprinted with permission from the January 1980 *Reader's Digest.* Copyright © 1980 by The Reader's Digest Assn., Inc.

plied with oil from California and Canada. What's more, the region lacked refinery capacity to handle the Alaskan crude. Since Alaska's oil was really needed in the Midwest, the sensible scheme would be to pipe it across Canada directly to Chicago. Why did the companies favor the Pacific route? They planned to export substantial quantities of the oil to Japan, said the analysis, where it would bring higher profits.

But the companies said the West Coast *did* need the oil and *had* sufficient refinery capacity for it. So, in 1973, Congress okayed the trans-Alaska pipeline. Three years later, as the project neared completion, the companies began pressuring the Carter Administration for permission to export the oil to Japan. *They said the West Coast had neither sufficient demand for it nor sufficient refinery capacity to handle it.*

What caused this reversal? The companies now say their original plans were upset by the Arab oil embargo and by environmental restrictions on refinery construction. But neither Congress nor DOE has looked into the facts behind this claim. So today, all the country knows for sure about Congress's decision is that it caused Alaska's oil to be delivered to the wrong side of the country.

☐ During the natural-gas shortage of 1977, gas producers pointed to official statistics showing an alarming long-term decline in US gas reserves. Deregulate the price of gas, they said, and they would have the incentive to explore new fields.

But industry critics disagreed. First, they said, the statistics showing a long-term decline in reserves could well be phony; they were the industry's own figures, drawn up in secrecy by the American Gas Association (AGA). As to the immediate shortage, critics said, the industry itself was causing it by holding gas underground in the hope of getting higher prices under deregulation.

In 1976 a Congressional subcommittee, deciding to look into the figures for itself, subpoenaed AGA's files and took sworn testimony from gas-company geologists. Under close questioning, they conceded that they were denied access to critical company data. One gas field which eluded the reserves estimate was so large that it would have increased AGA's tally of new US discoveries by 10 percent. One set of subpoenaed documents indicated that gas reserves in the Gulf of Mexico might be 60 percent higher than AGA's figures showed.

Nevertheless, in 1978 Congress deregulated natural gas. A few months later the AGA's chairman proclaimed that the long-term shortage was over. This continent, he wrote, had "an almost inexhaustible supply." What about the short-term shortage of the previous winter? "What we had then was not so much a shortage as a government blockage," he wrote. "The gas was there, but it stayed in the ground."

So skeptical of the industry's data has Washington become that two years ago DOE created an Office of Energy Information Validation solely to verify industry statistics. In July, I asked Charles Smith, the office's head, how many audits he had made of industry data. He cited six studies done by an outside contractor, the Oak Ridge National Laboratory.

I then questioned the man in charge of the Oak Ridge job, Andrew Loebl. He said he had not audited anything; he had merely checked on whether DOE's questionnaires were asking the right questions. "Charlie [Smith] expressly told us that audits are done by his staff, not by outside contractors," said Loebl.

Then is *anyone* in DOE's Energy Information Administration [EIA] doing any auditing? Says its chief, Lincoln Moses, "We are not funded to make actual audits."

So, after two years on the job, Moses' auditors have still made no audits.

Despite this, some observers maintain their confidence in the industry's data. The New York *Times*, claiming the figures "have revealed no large errors," attributes public doubts to "paranoia." However, one finding in the Oak Ridge study casts doubt on such confidence. In an informal poll, industry executives were asked about the accuracy of the forecasts of refinery output that they gave DOE. They admitted the figures were from 5 to 200 percent off base. The reason: deliberate distortion. The figures were "blue sky projections . . . highly influenced by competitive gaming and market posturing."

What's more, there is evidence that DOE is deliberately

helping the industry to hide facts. Last summer [1979], the Federal Trade Commission told Congress that DOE had frequently refused to give it vital industry data. Looking into the complaint, a House subcommittee found the refusal violated a 1974 information-collecting law that explicitly orders the agency to give federal law-enforcement agencies whatever data they ask for.

How can DOE get away with this? There's another information-collecting law that implies—but does not mandate—DOE should share the information. EIA chief Moses admitted that he uses the second law to get around the first.

Last May [1979], President Carter ordered DOE to conduct a vigorous investigation of the causes of the summer gasoline shortage. In July, DOE reported to the White House that it had found no wrongdoing by the industry. However, news reports subsequently revealed that DOE's "investigation" had: depended mainly on industry-supplied information; taken no sworn testimony, subpoenaed no data and conducted no audits of company books.

When I asked former Energy Secretary James Schlesinger about this, he said he had had all the information he needed without an investigation. Anyone who thought the companies had manipulated the shortage, he indicated, was an "oddball with an ax to grind."

Congress and the President should immediately:

1. Replace DOE's data collectors with officials who can be trusted to conduct rigorous audits.

2. Order the Energy Information Administration to obey the present law by sharing its data with law-enforcement agencies.

3. Undertake a Congressional investigation of all important cases in which the industry may have given the Government false information—including . . . [the 1979] summer's gasoline shortage.

These actions are urgently needed. As the matter now sits, a dozen or so private companies exercise almost total control over what facts the country is allowed to see in determining its energy future.

II. INVENTING SOLUTIONS

EDITOR'S INTRODUCTION

The energy debate heats up when participants move from diagnosing the crisis to suggesting specific remedies for it. Thomas C. Schelling, an economist, has defined the goal of such remedies: "The problem now," he said, "is to meet the rising economic cost of fuel with policies that minimize the burdens, allocate them equitably, avoid disruptions in the economy, and keep the cost from rising more than necessary."

That's a tough order, especially in view of the fact that most remedies have side effects that are more harmful to some patients than to others. People who supply the nation's energy consume energy, too, but their recommendations for a solution to the crisis are more likely to reflect the needs of the institutions they serve than their personal interests as consumers. The consuming public, on the other hand, is quick to condemn the suppliers for their roles in the energy crisis but much slower to accept the need for sacrifices in its own use of energy.

The Government—the President and his Department of Energy, and the Congress—is caught in the middle, trying to meet the challenge by setting a course acceptable to the majority. Such a goal, while not entirely out of reach, is not one to encourage boldness on the part of politicians. President Jimmy Carter proposed his first National Energy Plan in 1977. Congress did its best to ignore the President's proposals. In the end, it passed a dismembered program, and the energy crisis—and the debate—continued unchecked.

In the course of the debate there emerged a notion that has shaped informed opinion perhaps more than any other— namely, that there are two main ways to meet the challenge of America's energy future. As perceived by Amory B. Lovins, consultant physicist and British representative of Friends of the Earth, Inc., one choice is to take the "hard path"—meeting our energy demand with electricity generated largely from coal and nuclear fuels. The other, which he advocates, is the "soft path"—gradually phasing out the large generating facilities and replacing them with what he feels are more efficient, smaller units that rely on solar power.

In the selection that opens this section, Alexander B. Trowbridge, vice chairman of the Allied Chemical Corporation, speaking at a conference called to address Lovins' thesis, argues for a mix of the "hard path" approach and the "soft path" approach to the problem created by the shortage of cheap energy.

On April 5, 1979, President Carter tried to rally Congress behind a simplified program. The twin centerpieces of this program were a lifting of price ceilings on crude oil produced in the United States and a tax on roughly half the profits oil companies were expected to reap from decontrol. The second piece in this section, an article from *Time*, describes that phase of President Carter's renewed attack on the energy crisis. The President did not need the approval of Congress to lift price controls on domestic crude; it was within his power to let them expire, one part at a time, until all controls were gone by October 1, 1981—and that is what he is doing. Nonetheless, his proposal met, and continues to meet, strong opposition from consumer groups and members of Congress. In the third article, US Senator Howard Metzenbaum, a Democrat from Ohio, details the reasons for this opposition to oil price decontrol.

In the next selection—excerpts taken from two speeches made in July 1979—President Carter presents proposals for an "energy-secure America," including the creation of an Energy Security Corporation and the development of alternative sources of energy. Among other results, these speeches moved the debate over the nation's energy future to the question of "synthetic fuels"—liquids and gases made from coal, oil shale, and tar sands.

Next, Philip Shabecoff, a New York *Times* reporter writing in *Technology Review*, explains how politics gave this proposal life and how political considerations may kill it too. (Section V of this volume will look beyond the political argument to the feasibility of synfuel production.)

The section's final piece, an excerpt from the conclusion of the report of Harvard Business School's Energy Project, returns to Trowbridge's theme (expressed in the opening selection) and carries the idea further. The editors of the report, Robert Stobaugh and Daniel Yergin, call for a "balanced energy program," with an emphasis on conservation and solar power. Members of the Energy Project favor a "free market" approach to the problem, with the government encouraging conservation and a transition to solar power through tax incentives.

THE NEED FOR LARGE-SCALE ENERGY TECHNOLOGY[1]

Although the word *dichotomy* is commonly used to mean a forking into two parts, the dictionary also carries a definition pertaining to *logic* which reads, "a division of a class into two opposed subclasses, as in *real* and *unreal*." I think that definition hits the energy debate right on the button.

We have been told that one path is contesting with the other, and a choice is in order; a mutually exclusive choice at that. We have been served notice that the traditional hard path, which means big technology and central station power generation, is a dead end road. The soft path is presented as virtuous, benign, economical, life affirming and in general more fun to have around.

While the hard path involves primary reliance upon expansion of centralized high technologies to increase supplies of energy, especially electric power, the soft path emphasizes conservation and increased efficiency, widespread employment of renewable energy sources, and strict emphasis on

[1] Excerpts from the address "The Systems Approach to Energy," by Alexander B. Trowbridge, vice chairman, Allied Chemical Corporation. *Vital Speeches of the Day.* 45:764–8. O. 1, '79. Reprinted by permission.

matching scale and energy quality to end use needs. It also accepts a transitional and ever diminishing role for certain fossil fuel technologies.

In concept there is little in the soft path route which would seem either unreasonable or imprudent except for the questions of time and of near-term economics. Indeed, common sense tells us that special emphasis should be placed upon such approaches as a supplement to the traditional hard path, larger technologies. . . .

[Consultant physicist Amory B.] Lovins warned in his famous October 1976 *Foreign Affairs* article that "It is important to recognize that the two paths are mutually exclusive. Because commitments to the first may foreclose the second, we must choose one or the other—before failure to stop nuclear proliferation has foreclosed both."

And thus the lines of the two paths have been harshly drawn for us. The leading edge of this controversy, of course, is nuclear power, and upon that technology descends most of the rational and irrational debate. According to the now conventional wisdom of soft path advocacy, all scenarios which include any variation on the use of nuclear power are completely unacceptable. The major premise seems to be that nuclear power plants in themselves are an almost automatic invitation to wholesale manufacture and proliferation of nuclear weapons.

This argument fails to deal with the simple truth that no nation now possessing nuclear weapons has used a conventional power reactor to make them. And further, there are well-documented easier, cheaper and less risky ways to get the basic stuff for nuclear weapons than through a nuclear power program. It's been shown that if all nuclear power plants were eliminated, there would still be the same amount of risk and potential for various nations to decide to build nuclear weapons. Nuclear power and nuclear weapons share the word nuclear, but having one does not beget the other unless you consciously equip yourself to extract plutonium and purposefully use it in a weapons program.

If fully utilized, this nuclear fuel cycle, which would include breeder reactors, is capable of providing fuel for electric power likely in excess of all the known domestic coal reserves in this nation. It is a massive energy resource waiting to be utilized. And yes, there are risks associated with full use of the nuclear fuel cycle and, at the same time, formidable risks which result from *not* using our nuclear resources. Most of these nuclear-related risks can be virtually eliminated; others can be minimized and others will remain. There is no "risk free society," and no real point in assuming it is possible to attain one.

Nevertheless, the equivalent of the death penalty has been suggested for nuclear and other large-scale energy technology. One curious justification is that they are "inelegant," because of a mismatch between the large scale of the major generation facilities which often produce energy for rather modest end use tasks. This has been likened to using a chain saw to cut butter. This fine imagery works both ways, for industry would indeed find it difficult to shape steel with a paring knife.

It has been written that "the most attractive political features of soft technologies and conservation . . . may be that, like motherhood, everyone is in favor of them." We should also note that motherhood is not what it used to be, and issues like zero population growth and abortion are hardly benign political issues these days. As the costs, impacts and ramifications of soft technologies become more evident, it's likely that the elite favoritism which now is so buoyant to them will give way to the same kind of controversy which surrounds every other choice and decision facing us in the energy area.

I cannot accept the existence of or the need for a soft versus hard energy paths dichotomy. The theories advanced in support of such a separation of powers are unconvincing and counterproductive. The most glaring errors in such a point of view are, in my judgment, as follows:

(1) There is no clear evidence that an exclusive choice is necessary between large and small centralized or decentralized energy courses. Quite to the contrary, there is a clear and definite need to combine and pursue both.

(2) Assumptions for gains to be had from conservation, increased efficiency of energy use, solar power and other soft path technologies have been enthusiastically promoted without consideration by the public of the possibility that they cannot achieve the gains forecast.

(3) Back-up energy supplies will be necessary and indeed called upon for most soft path technologies and these will still call for large-scale energy facilities.

(4) Most of the public still fails to recognize that solar power's greatest proven attribute is for comfort heating for hot water mainly—and that means saving mostly oil and natural gas which are now primarily used for comfort heating. While solar power, insulation and conservation *can* achieve a significant savings in many areas of the country for water and space heating, that will not make a significant dent in the need for electric power, and will make only a modest but not insignificant savings in the overall energy picture. Transportation is what consumes most of our energy. Solar power, insulation, conservation, windmills and other soft energy paths will not power our trucks and autos. Hard path big-scale technology can provide the large amounts of electric power for more efficient mass transportation and for widespread deployment of the electric vehicle for family use, providing we can solve other related problems such as battery performance. Again, the *exclusive* use of a one-way soft path is a detour which need not be made.

(5) Those who are most adamant about cutting our ties with big technology often insist that business and industry are against solar power, because they can't profit from the sun. They insist solar energy is free and, as such, this frightens the "greedy businessman." Again, while this may appeal to the current prejudices, consider that full solar development, deployment, servicing and such will mean billions of dollars in sellable items and services. No force has yet been able to withstand development of a new and profitable item demanded in the marketplace. Quite the contrary, there is a rush to compete for the market. With world energy prices soaring, previously uneconomic energy system dreams of

countless past inventors and enthusiasts are only now becoming economically feasible. It will take a healthy economy, substantial money for research, development and manufacturing, and substantial amounts of hard-path-generated energy to fabricate and supply soft path energy systems.

(6) It is said that exclusive allegiance to the soft path will result in more jobs than the hard path. Perhaps so, if you consider just the statistics. But what the statistics mean is that we will have to completely turn our organized labor force upside down and retrain and relocate vast numbers of workers. Such changes, if called for, can only be accommodated gradually, for you are talking about people and professions and a way of life they have chosen—not been assigned. If the soft path creates a vast new potential for employment, all the better. But do not expect labor to take this promise on faith alone.

(7) Finally, there is the matter of using the soft path as a social engineering tool. . . . Lovins, Dr. Barry Commoner, Ralph Nader, Tom Hayden and others not only have used the word *freedom* to describe energy independence but have used it in terms of glowing social and economic restructuring. But every kind of economic and political system around the world has taken the same approach by using big technology, and these governments are also committed to conservation and other small-scale technologies. Lenin promptly launched the Soviet Union down the path of electrification and big technology. And decades later Mao took the soft path route. Today the USSR still holds to the hard path and is also working hard on small technologies, while China has repudiated the exclusive road of Mao and the Gang of Four [Chiang Ch'ing (Mao's wife), Wang Hung-wen, Chang Ch'un-chi'ao, and Yao Wen-yuan, high-ranking radical supporters of Mao] and is combing the world for big technology. Nowhere has any nation of size taken or long endured a single, exclusive energy or technology path.

This debate over hard versus soft technologies and the accompanying antinuclear actions is taking place against a very real backdrop of dwindling energy reserves and supply.

As all of you know, regardless of what the estimates might

be, there is universal agreement that oil and natural gas are finite resources, and that if we continue to consume them at current rates we will exhaust them—even with newly discovered sources. Coal is also a finite resource, albeit with far greater reserves to count on. Companies such as mine, in the petrochemical industry, cannot exist without feedstocks made from these fossil fuels. Everything from fertilizer to fibers, aspirin and paints come from such petrochemicals, And rising oil prices drive up the cost of every product made from them.

Even as we sat in the long gas lines this summer [1979], we were warned of the possibility of home heating and diesel oil shortages for this winter. And long term forecasts by industry and the government show a growing and eventually enormous gap between electric power demand and supply in the late 1980s and beyond. A gap so huge that it cannot be filled by power plants now proposed or under construction, including many which are being delayed by controversy. Contributions from conservation and solar power are *not* going to be adequate to substantially change this bleak picture. At the same time we face energy shortages, we must also curtail and substantially cut back on our importation of oil from OPEC and other foreign sources.

A new report by the Energy Project at the Harvard Business School reaches the conclusion that the only viable program that would politically reduce US dependence on imported oil would be for the government to offer enormous financial incentives to encourage conservation and use of solar energy. This report, titled *Energy Future* [Random House, 1979], recommends subsidies for mass transit to cut down on gasoline demand; and massive additional subsidies of solar power—up to 50 percent of the installation costs. The authors see these government subsidies as incentives, and conclude that "the pursuit of profit has, after all, served American society well in the past, and clearly the carrot makes for better politics and more acceptable change than does the stick." [See "Needed: A Balanced Energy Program," below in this section.]

Of particular interest is this statement from the report:

A politically acceptable program that can make a significant contribution to a solution of the energy problem is better than one that might theoretically solve it altogether but which has no chance of being adopted.

The authors of this report conclude that although America needs gas, coal and nuclear power, because of controversy and political considerations it's unlikely that they can deliver vastly increased energy supplies.

In my view, they're both right and wrong. They're wrong when they say that we can't get vastly increased energy supplies from oil, gas, coal and nuclear power. The potential is certainly there. What's required is to cut back on government delays and the unfettered ability of various opposition groups to constantly hold up many of these projects. The President's Energy Mobilization Board has been proposed to tackle this problem. If electric utilities can find it easier to burn coal, they too can produce substantially more electric power. Domestic on- and offshore oil resources can be substantially improved if some of the disincentives are removed. Nuclear power could easily shoulder an enormous burden of electric power production if the siting and licensing procedures were shortened instead of eternally lengthened. And, we've already seen that financial incentives have resulted in far greater exploration for natural gas in a relatively short period of time.

But, at the same time, the authors are right when they observe that the key to these and other technologies is in their political acceptability.

A recent statement by a staff lawyer for the Natural Resources Defense Council blasted President Carter for even suggesting that energy development and environmental protection were not legitimate and implacable foes. Said Mr. Richard Ayers on August 5th [1979], "By denying the legitimacy of the conflict between energy development and the environment, the President's proposals not only were terrible in and of themselves, they were also a serious blow at the environmental movement." We're going to have to find a basis for reducing this kind of polarization and to recognize that consumers and Americans concerned about their environment are not two constituencies, but one. . . .

Perhaps our greatest problem is not only to reconcile energy supply and demand, but to develop and implement a systems approach which will cope with and reduce controversy. . . . In my opinion, the notion that the so-called soft path can take the place of large-scale technology is both wrong and naive, perhaps even dangerous to the best interests of all of us. A partnership between the technologies makes the best sense, and it's clear that all the approaches are going to need all the help they can get. . . . One need not starve or eliminate large-scale hard technology in order to encourage the growth of decentralized soft technologies. Synthetic fuel conversion plants, coal and nuclear electric generating stations, large solar electric power plants, hydroelectric projects, all have their place and their contribution to make in a prudent, systematic plan to diversify our energy supply and its management.

I'd like to sum up my views and opinions on the matter of hard versus soft energy paths, the controversy over energy and the prospects for use of a systems approach.

(1) I see no dichotomy between the so-called soft and hard energy paths; quite the contrary, they should be seen as being capable of delivering differing quantities of energy in differing time periods. They should be selected and implemented in the order in which they can adequately meet our needs.

(2) The merits of a strong effort at conservation are obvious and should be vigorously pursued. Energy efficiency measures the energy required to produce a unit of output. Improvements in energy efficiency represent conservation in its truest sense—making energy do more, or go further. Due to improved operating techniques, investments in energy-saving equipment and the application of new, energy-efficient technologies, industry's energy efficiency has improved 16 percent since 1973. . . . At the same time, this effort should not blind us to the need to pay more immediate attention to resolving the obstacles blocking an increased domestic energy

supply. Toward that end, we must utilize the full spectrum of domestic energy resources, including coal, natural gas, oil and nuclear power.

(3) Americans have always considered controversy as a healthy and productive exercise. While the right to dissent and controversy should be preserved and protected, some means must be found to enhance consensus views and to finally permit actions to be taken in the overall public interest. We need to develop a conscious effort towards depolarization of views. . . .

(4) Systems analysis is an extremely useful and valuable tool for planning a complicated and concerted program with many component parts, and for understanding sensitivities and ramifications.

While I have my doubts about applying it as a tool and technique to resolve controversy, I will listen to any proposal to help develop such an approach or program. . . . We lack neither the resources nor the talent to solve virtually all of the problems which have come to perplex us in considering what to do about energy supply and demand.

PHASE 1 OF CARTER'S ENERGY PLAN: DECONTROL[2]

This is a painful step, and I'll give it to you straight. Each one of us will have to use less oil and pay more for it.—Jimmy Carter, Address April 5, 1979.

Straight it was. When he announced his first energy policy, way back in 1977, Jimmy Carter summoned the nation to a "moral equivalent of war," which was to be fought through a highly complex program of tax incentives and other gimmicks, and focused on conservation as the key to solving the nation's twin problems of declining oil production and rising dependence on price-gouging foreign suppliers. The new

[2] Excerpts from the staff-written article, " 'Use Less, Pay More.' " *Time*. 113:66–8. Ap. 16, '79. Reprinted by permission from *Time*, the Weekly Newsmagazine; copyright Time Inc. 1979.

plan that he outlined in his plain-spoken twenty-three-minute Oval Office addresss . . . [on April 5, 1979] was far simpler—and much more likely to be effective. Henceforth, old-fashioned marketplace economics is to be the basic engine to spur not only fuel saving but also a much needed, intensified search for new domestic supplies. But as Carter promised, the change will be painful: during the coming months and years, US oil prices will leap up, forcing consumers to dig even deeper into their pockets to pay for gasoline and heating oil and giving an upward kick to the country's already hurtful inflation.

The new policy also promises much political pain and peril for Carter. The essence of his program is to strip away the controls that have held the cost of domestically produced crude oil at artificially low levels ever since the postembargo days of 1974. Next month, using Executive authority, he will order a gradual phase-out of the controls so that they will be entirely eliminated by October 1, 1981, when by law they would have expired anyway.

To prevent handing what he sees as an unearned bonanza to the oil companies, Carter called on Congress to enact a "windfall profits tax." It would skim off about half the $13 billion or so of extra revenue that oil firms stand to get as the price of domestic crude oil, which now averages $9.45 per bbl., rises to the world level. . . . Under Carter's plan, the proceeds of the oil tax would be funneled into an Energy Security Fund that would bankroll the development of alternative energy sources such as solar power and coal gasification, help low-income families pay for the rising cost of fuel and stimulate the development of energy-efficient mass transit systems such as rail and bus service. . . .

In blunt terms, the President sought to dispel the notion, reflected in polls, that most Americans feel the oil problem is somehow phony. "The energy crisis is real," he emphasized. The nation's dependence on foreign oil, which now supplies nearly 50 percent of the United States's needs, up from 36 percent in 1973, has left the country gravely vulnerable. As the President said, "Our national strength is dangerously dependent on a thin line of oil tankers stretching halfway around the earth, originating in the Middle East and the Persian Gulf—one of the most unstable regions in the world."

In his message, Carter announced a cluster of measures—some substantial, others symbolic—to help deal with the energy situation. Among them:

☐ *Parking restrictions.* To discourage the use of cars for commuting to and from work, Carter said that he would eliminate free parking privileges for federal employees nationwide. He urged private corporations to do the same.

☐ *State allocations.* As a further move to curb gasoline demand, which is rising almost three times as rapidly as oil consumption as a whole, Carter announced a plan to bring state governments into the conservation act. He said that he would soon set strict gasoline reduction timetables for all fifty states, and that if they were not met he would ask for mandatory weekend closings of service stations.

☐ *Voluntary driving cuts.* The President asked each of the nation's 138 million licensed motorists to drive fifteen miles a week less than they do now. The fuel savings could total 413,-000 bbl. of oil every day. That is nearly half the amount of oil consumption that the United States pledged to cut during 1979 as part of a coordinated conservation drive by the nineteen member-nations of the International Energy Agency.

☐ *Red-tape reductions.* To make it easier for important new energy projects such as refineries and pipelines to come onstream without years of delays, regulatory hearings and appeals, Carter signed an Executive order setting strict deadlines for processing applications. He also said that the Administration would take action to slice through the bureaucratic barriers that have bogged down plans by Standard Oil of Ohio for a pipeline to carry Alaskan oil from California to Texas. The pipeline would enable some 350,000 bbl. per day of Alaskan oil to reach Eastern markets, thereby displacing the need for an equal amount of imports.

Yet the key to Carter's program is crude-oil decontrol. From the moment that President Nixon set up controls in December 1973 to prevent the price of US oil from chasing

OPEC crude into orbit, the whole cumbersome apparatus has provoked one wrangle after another between the oil industry, Congress and the White House. Just as Carter is now doing, Nixon and Gerald Ford also tried to dismantle price controls on domestic oil and tax away the resulting profits. All that those Presidents accomplished was to get themselves caught in crossfire quarrels between oil industry demands for immediate and full decontrol and equally insistent counterarguments from consumer groups and the industry's many critics on Capitol Hill.

Under Carter, the struggle over price controls has already produced some bitter contention, and that is now certain to intensify. In fact, the single most surprising aspect of Carter's entire message was its harsh indictment of the oil industry. More than just populist politics with a dash of down-home demagogy, the President's assault was a bold—perhaps even slightly desperate—gamble to outflank industry lobbying efforts in Congress and rally public opinion behind the profits tax. Carter's aides openly concede that this is White House strategy. Says one presidential assistant: "We're on the side of the angels this time, for once."

Carter all but accused oilmen of energy treason in the name of profit, and he appealed to the nation to deluge Congress with demands that it pass his tax if for no other reason than to stop the oil firms from benefiting from the public's energy problems. Said the President: "Just as surely as the sun will rise, the oil companies can be expected to fight to keep the profits that they have not earned. Unless you speak out, they will have more influence on Congress than you do."

Carter also sought to put Congress on the spot, saying,

Please let your Senators and Representatives know that you support the windfall profits tax, and that you do not want the need to produce more energy to be turned into an excuse to cheat the public and damage our nation. Every vote in Congress for this fund will be a vote for America's future, and every vote against it will be a vote for excessive oil company profits and for reliance on the whims of the foreign oil cartel.

Though oilmen applauded decontrol, none welcomed the

abuse they were getting from the President. Complained a Mobil vice president: "I have no idea why the President thinks it is in the interests of the United States to pit citizens against citizens. The way he was talking he seemed to think oil companies were a foreign country." Added T. Boone Pickens Jr., president of Mesa Petroleum, a big Texas-based exploration and drilling firm: "Why these frontal attacks on the industry? Never once does he say what a fine job we've done in finding 90 percent of the oil and gas that there is in the world. Instead, he almost infers dishonesty." One of the cooler judgments came from Thornton F. Bradshaw, president of Atlantic Richfield. Said he: "I'm not altogether sure that this kind of tax is necessary, but if Congress thinks so, then that's what will happen. If it is politically essential to accomplish decontrol, so be it." . . .

[Eight months later, in December 1979, both houses of Congress passed "windfall profits" bills. In January 1980, the two houses agreed on a complicated formula that would extract $227.3 billion in windfall profits taxes during the 1980s, the bulk of them from the major oil companies. The bill was signed into law in April 1980.—Ed.]

Carter has also called on Congress to revoke one of the oil industry's most zealously protected overseas business perks: the foreign tax credit. Such credits are earned when US companies pay income taxes to foreign governments. To prevent double taxation, US law permits the payments to be used to reduce, on a dollar for dollar basis, the amount of income tax that a company must pay to the IRS. [As of April 1980, there has been no change.—Ed.]

Big companies have reaped large benefits by structuring their payments to foreign oil nations to enable levies that would normally be considered royalty payments for the purchase of crude to qualify as income taxes. More than 75 percent of all foreign tax credits claimed by US companies now go to oil and gas firms. . . . [In 1978] the companies saved an estimated $1.2 billion or more on taxes that they would otherwise have had to pay to the IRS.

Whatever the fate of the tax, decontrol alone will still

bring benefits. Not only will rising prices, which are expected by the Administration to push up the cost of gasoline by 5¢ to 7¢ per gal. by 1982, encourage people to waste less fuel, but increased revenues to oil companies will certainly give the industry the financing needed to boost drilling activity.

There are, of course, uncertainties and risks. Though economists are willing enough to guess, none can say with confidence what the ultimate inflationary impact of decontrol will be. Nor is it entirely clear just how much decontrol will increase domestic oil production. By Administration reckoning, the gradual phase-out of controls should encourage companies to pump more and more oil from their wells until, by 1982, production reaches an additional 700,000 to 800,000 bbl. daily (the United States now uses about 19 million bbl. per day). That would displace an equivalent amount of imported oil, but energy demand throughout the economy would itself be growing. In effect, increased production from existing wells would do little more than keep pace with rising imports.

To displace significant amounts of imports, huge new oil-fields will have to be discovered and developed. Unfortunately, the oil may just not be there to find. Even though oil companies drilled more than 48,000 new wells around the nation ... [in 1978], nearly double the amount of 1973, production continues to decline gently but steadily. A new crash program of drilling could turn out to be a multibillion-dollar disappointment.

For all that, decontrol remains the most effective energy policy step that the President is able to take. By allowing domestic crude prices to rise to world levels, Carter has sent a clear signal to the nation's trading partners and allies that the United States is at long last beginning to face up to the difficult decisions forced upon it by the energy squeeze. [For a statement of the opposition, see "The Case Against Oil Price Decontrol," by H. M. Metzenbaum, the following article in this section.—Ed.]

THE CASE AGAINST OIL PRICE DECONTROL[3]

Decontrol is a cruel policy. It is a policy that imposes disproportionate sacrifices upon those who can least afford them; it fuels the inflation that is steadily eroding the economic gains that the working men and women of this nation have made in recent years; and it surrenders the last vestiges of independence that this country has from the pricing decisions of the OPEC [Organization of Petroleum Exporting Countries] cartel.

With decontrol in effect, every drop of oil in this country will sell at the artificial price established by OPEC. We will have locked the fundamental economic interests of our own oil producers into the OPEC structure. Moreover, we cannot overlook the fact that the higher the OPEC prices go, the higher the profit for the oil companies.

Exxon's profits for the first quarter of 1979 were up by 37 percent over the same period last year. Gulf was up by 61 percent, Amerada-Hess 242 percent, and Sohio 303 percent. Texaco, whose board chairman stated recently that his company is not making enough, had first-quarter profits that were up by 81 percent over the first quarter of 1978—and a truly staggering 342 percent over the same period in 1977.

This is only the beginning. Crude oil decontrol will produce over $21 billion in new revenues for the oil companies in the next twenty-eight months. Between now and 1990, the oil companies will take in $58 billion from decontrol of upper-tier oil alone.

Despite this, a recent study by the Congressional Budget Office concluded that those billions will bring to the people of this country no appreciable growth whatever in oil production. For that matter, the entire crude oil decontrol program will bring in approximately 200,000 new barrels of oil per

[3] Excerpts from the article by Senator Howard M. Metzenbaum (Democrat, Ohio). *USA Today.* 108:18–19. S. '79. Reprinted by permission. Copyright © 1979 by the Society for the Advancement of Education.

day—a drop in the bucket when compared with the 20 million barrels a day that we now consume.

The oil companies do not need more incentives to produce in this country—they already have the most generous incentives in the world. Yet, in spite of those incentives, production in the Lower 48 States continues to fall. The oil companies say that they need high profits to restore domestic production. They say that they need money to bring on line new sources like shale oil and tar sands and heavy oils. However, the record shows that, too often, the industry does not use its profits to increase energy production. The record shows that, time and again, the major oil companies use their earnings to expand in areas that are wholly unrelated to gas and oil.

In 1977, ARCO bought Anaconda Copper. Tenneco has gone into Holiday Inns, almond orchards, auto equipment, and life insurance. In 1976, Mobil acquired the Montgomery Ward Department Store chain and the Container Corporation of America. In that same year, Mobil made a $520 million bid for a huge real estate deal in Southern California.

Exxon has expanded into business machines and word processing equipment. Tomorrow morning, if it chose to do so, Exxon could use its cash on hand and liquid assets to buy out J. C. Penney, Dupont, Goodyear, and Anheuser-Busch—without going one penny into debt.

Windfall Profits Tax

President Carter has said that he wants a windfall profits tax. The President has said that he is against the kind of windfall profits tax—the so-called plowback provision—that would take money from the oil companies with one hand and give it back to them with the other.

We have heard the strong words from the President about preventing the oil companies from ripping off the American people, but the windfall tax proposal made by the Administration is only a shadow of those tough words. By making the tax a deductible business expense, the so-called 50 percent windfall profits tax will be reduced to a very modest level—

less than 12 percent at its peak in 1982 on extra profits of some $14.5 billion.

The oil industry has been strangely silent about imposition of this tax. No wonder.

The President's windfall tax proposal is a farce. . . . If the President is truly outraged about windfall profits, there is one sure and certain way to avoid them. That way is to stop the windfall in the first place.

In June 1975, at a time when President Ford was advocating oil price decontrol, a distinguished American made an eloquent statement in opposition to that proposal:

If the Gerald Ford–oil industry policy is implemented: it will add from three per cent to four per cent to the nation's inflation rate; it will cost us consumers more than $30,000,000,000 annually, draining this purchasing power away from other parts of the floundering economy and increasing already disgraceful levels of unemployment; it will encourage additional OPEC oil price hikes; it will not result in decreased consumption equivalent to price increases because of inelastic demand for certain petroleum products; it will punish those with low and middle incomes, while the rich continue to waste all the fuel they want; it will continue a callous disregard for environmental quality. In short, the Ford–oil industry energy policy is merely another example of letting the average American pay for the politicians' mistakes.

Jimmy Carter was right when he said that in 1975. Decontrol was a mistake in 1975 and it is a mistake in 1979.

Decontrol is a dead end. It has *not* worked to increase the supply of natural gas, and it will not work for crude oil either. It will strengthen OPEC, it will massively boost inflation, but it will do nothing—absolutely nothing—to lessen this country's dependence on foreign oil.

Decontrol will do nothing—absolutely nothing—to restore stability to our economy. It will do nothing to rally the American people behind a fair and balanced national energy policy. Decontrol is a policy that abandons even the hope that our government can determine the energy future of this nation.

Yet, the Department of Energy would like the people of this country to believe that the Administration is getting tough on the oil industry. Recently, the DOE announced action aimed at recovering $1.7 billion in excess oil company charges to consumers.

At first glance, that looks impressive. It sounds tough, but, in fact, it is not. Two years ago, the Securities and Exchange Commission reported that overcharges at that time had already reached many billions of dollars. The DOE has done nothing about it until now, and what has been done is far too little, far too late, and far too obviously timed to convey the impression that DOE is a vigilant defender of the public interest.

A recent study of DOE by Common Cause, the citizen lobby, gives a very different picture of the true relationship between the industry and [former Energy Secretary] James Schlesinger's DOE. According to Common Cause, in DOE's first year, top officials met with industry representatives five times more often than with consumer or public interest representatives.

Common Cause reports that industry representation on DOE advisory committees is six times greater than representation of the general public. The study further points out that DOE is the government-wide champion in rejecting freedom of information requests from the public. Fully 60 percent of the requests are rejected in whole or in part. I believe that this is unacceptable, and I believe that the people of this country have a right to expect more from their government.

What Can We Do?

I am sure that many of you are thinking to yourselves at this point that it is well and good for Howard Metzenbaum to criticize the Administration's proposals, but what does *he* propose?

I don't say that I know all the answers. I don't even say that I know most of them. But I do know some of them.

Since the Nixon Administration's Project Independence in 1974, we have tried a series of voluntary measures to conserve energy. Those measures are not strong enough. I believe that this nation must move toward mandatory energy conservation. The word *mandatory* sounds forbidding, and that's what it should be—forbidding!

We in this country should be forbidding the manufacture and importation of automobiles that do not meet minimum mileage standards, but we are not doing that. We should be doing everything possible to bring the inventive genius of this country to bear on producing new forms of energy. Yet, it is two years since the President's first energy message and the DOE still does not have a director for solar applications.

We should be retrofitting homes in this country with proven devices to increase home furnace efficiency. We should be requiring minimum efficiency standards for the thousands upon thousands of electric motors and pumps used by industry.

We should be requiring the states to enforce the 55-miles-per-hour speed limit. That alone would save more oil than decontrol will produce. We should be working on a rush basis to liquify and gasify our vast reserves of coal. The Germans liquified coal on a large scale during World War II. South Africa is doing it right now. If these countries can do it, I don't know what is holding back the United States.

This country needs competition in its energy industry. To get it, we should require the oil companies to divest their massive holdings in coal and uranium and solar patents. We should require true competition between oil and coal and solar and nuclear power. However, we are not doing that.

If we really want to challenge OPEC, we should be extending vigorous assistance to developing countries around the world that want to drill for oil. It is a fact that more oil wells have been drilled in the state of Kansas than in all of South America. It is a fact that huge areas of the world have never been explored for oil and gas. It is also a fact that, recently, the president of our largest oil company registered a formal protest against a modest plan by the World Bank to help developing countries drill for oil.

We should be helping them. We should be doing all of these things that I have mentioned. We should be doing much more. Above all, we should be doing these things with the sense of urgency that the challenge before us demands.

PHASE 2 OF CARTER'S ENERGY PLAN: SYNFUELS AND ENERGY SECURITY

Synfuels[4]

The energy crisis is real. It is worldwide. It is a clear and present danger to our nation. These are facts and we simply must face them. What I have to say to you now about energy is simple and vitally important.

Point 1: I am tonight setting a clear goal for the energy policy of the United States. Beginning this moment, this nation will never use more foreign oil than we did in 1977. Never. From now on every new addition to our demand for energy will be met from our own production and our own conservation.

The generation-long growth in our dependence on foreign oil will be stopped dead in its tracks right now.

And then reverse as we move to the 1980s. For I am tonight setting the further goal of cutting our dependence on foreign oil by one half by the end of the next decade—a saving of over 4.5 million barrels of imported oil per day.

Point 2: To insure that we meet these targets, I will use my presidential authority to set import quotas. I am announcing tonight that for 1979 and 1980 I will forbid the entry into this country of one drop of foreign oil more than these goals allow. These quotas will insure a reduction in imports. . . .

Point 3: To give us energy security, I am asking for the most massive peacetime commitment of funds and resources in our nation's history to develop America's own alternative sources of fuel from coal, from oil shale, from plant products

[4] Excerpted from the address entitled "Energy Problems; the Erosion of Confidence," delivered on July 15, 1979, by President Jimmy Carter. *Vital Speeches of the Day*. 45:644–5. Ag. 15, '79. Reprinted by permission.

for gasohol, from unconventional gas, from the sun. I propose the creation of an Energy Security Corporation to lead this effort to replace 2.5 million barrels of imported oil per day by 1990. The corporation will issue up to $5 billion in energy bonds, and I especially want them to be in small denominations so that average Americans can invest directly in America's energy security.

Just as a similar synthetic rubber corporation helped us win World War II, so will we mobilize American determination and ability to win the energy war. Moreover, I will soon submit legislation to Congress calling for the creation of this nation's first solar bank, which will help us achieve the crucial goal of 20 percent of our energy coming from solar power by the year 2000.

These efforts will cost money, a lot of money. And that is why Congress must enact the windfall profits tax without delay. It will be money well spent. Unlike the billions of dollars we shift to foreign countries to pay for foreign oil, these funds will be paid by Americans to Americans. These funds will go to fight, not to increase, inflation and unemployment.

Point 4: I'm asking Congress to mandate—to require as a matter of law—that our nation's utility companies cut their massive use of oil by 50 percent within the next decade and switch to other fuels, especially coal, our most abundant energy source.

Point 5: To make absolutely certain that nothing stands in the way of achieving these goals, I'll urge Congress to create an energy mobilization board which, like the War Production Board in World War II, will have the responsibility and authority to cut through the red tape, the delay and the endless roadblocks to completing key energy projects.

We will protect our environment. But when this nation critically needs a refinery or pipeline, we will build it.

Point 6: I am proposing a bold conservation program to involve every state, county and city, and every average American in our energy battle. This effort will permit you to build conservation into your homes and your lives at a cost you can afford. I ask Congress to give me authority for mandatory conservation and for standby gasoline rationing. [In

October 1979, Congress gave the President the authority he sought to order gasoline rationing in an emergency.—Ed.]

To further conserve energy, I'm proposing tonight an extra $10 billion over the next decade to strengthen our public transportation systems. And I'm asking you, for your good and your nation's security, to take no unnecessary trips, to use car pools or public transportation whenever you can, to park your car one extra day per week, to obey the speed limit and to set your thermostats to save fuel. Every act of energy conservation like this is more than just common sense. I tell you it is an act of patriotism.

Our nation must be fair to the poorest among us so we will increase aid to needy Americans to cope with rising energy prices.

We often think of conservation only in terms of sacrifice. In fact it is the most painless and immediate way of rebuilding our nation's strength. Every gallon of oil each one of us saves is a new form of production that gives us more freedom, more confidence, that much more control over our own lives so that solutions to our energy crisis can also help us to conquer the crisis of the spirit in our country. It can rekindle our sense of unity, our confidence in the future, and give our nation and all of us individually a new sense of purpose.

You know we can do it. We have the natural resources. We have more oil in our oil shale alone than several Saudi Arabias. We have more coal than any nation on earth. We have the world's highest level of technology. We have the most skilled work force, with innovative genius.

And I firmly believe we have the national will to win this war.

Energy Security[5]

Last night I set forth a general strategy for winning the energy war—one which will enable us to meet all of

[5] Excerpted from the address entitled "An Energy Secure America," delivered before the Association of Counties, Kansas City, Missouri, on July 16, 1979 by President Jimmy Carter. *Vital Speeches of the Day.* 45:647–8. Ag. 15, '79. Reprinted by permission.

America's new energy needs from America's own energy resources. We will have to succeed, both by conservation and production, because from this time forth we will never import one drop of oil more than we did two years ago in 1977. I am drawing our line of defense here and now.

Now I'm going to mention one of the biggest figures you've ever heard—overall, we are going to make the unparalleled peacetime commitment—an investment of $140 billion for American energy security so that never again will our nation's independence be hostage to foreign oil.

Where is the money coming from? All of this investment of federal funds must come from the windfall profits tax on the oil industry which I have proposed to the Congress. It is now more critical than it ever was that Congress swiftly pass a strong, permanent windfall profits tax, and I want each of you, as county leaders, and all Americans who hear my voice, to bring your full power to bear to make sure that Congress acts to give the American people the financial weapon to win the energy war.

Let me brief you now on some of the new points of attack. Each year I will set targets for the amount of foreign oil we import. I am today announcing a quota—for this year, 1979—which will hold United States imports to 8.2 million barrels per day . . . 400,000 barrels per day below what we used in 1977. . . .

The Energy Security Corporation that I proposed last night to produce American energy from new sources will not be . . . just another federal agency. It will be outside the Federal Government—outside the federal bureaucracy—free to use its independent business judgment in order to produce enough alternate energy sources to meet its ten-year target of reducing our imports by 2.5 million barrels of oil per day. I am announcing new incentives for the production of heavy oil—oil shale and hard-to-get-at natural gas, all of which this country has in great abundance.

To make sure that energy projects such as critical pipelines, port facilities, production plants are built, a new Energy Mobilization Board will slash through red tape and bureaucratic obstacles and will set absolute deadlines for action at the federal, state and local level.

We are leaving with state and local authorities the first line of responsibility to remove roadblocks to these critical projects, but our energy crisis is so severe that if any level of government fails to act within a reasonable time, this board will see to it that action is taken, just as similar boards made sure that action was taken to protect our nation's existence in World War II. It's time for us to take this bold action, and we will.

I want to make energy goals as compelling to every homeowner and every renter as for business and industry. Utilities must shift from oil to coal, but I am also proposing to Congress a unique new program to require gas and electric utility companies to provide low-cost loans to their customers—the homeowners of America—to finance conservation improvements, repayable only at the time of sale of the home.

I am proposing another program that will offer incentives to convert homes that are now oil-heated to natural gas and to help oil-heated homes which cannot convert with conservation.

I've earmarked $60.5 billion in new funds for the next ten years to improve buses, subways and other mass transit and to build more fuel-efficient automobiles.

Recognizing that low-income families have been hardest hit by the OPEC price increases and rising energy costs in general, I am proposing to triple the size of the assistance program which I recommended to Congress on April 5. This will also be paid for out of the windfall profits tax which Congress is already considering.

In addition, we will be expanding the weatherization program to improve homes and to make them more efficient in conserving energy.

I want to explain one thing very clearly, because misinformation is being spread among the American people. I want you to listen to this.

We are working very closely with Mexico and Canada. The total quantity of production and export of oil and gas from Mexico is obviously a decision to be made by the people and the Government of Mexico. But we now purchase more than 80 percent of all the oil exported by Mexico—more than 80 percent. We are now negotiating a new agreement to purchase the natural gas which Mexico will be willing to export.

Although Canada's oil production will be steadily dropping during the next few years, we will continue to share hydroelectric power and other energy sources with our neighbor to the North.

One major project will be the new pipeline to be built from Alaska through Canada to bring natural gas to the Lower 48 States. By 1985 Alaskan and Canadian natural gas can displace almost 700,000 barrels of imported oil per day.

The North Slope producers have dragged their feet in helping to finance the pipeline needed to bring that gas to market. I've instructed the Secretary of Energy to call them in and get them going, and I will insist personally that this gas pipeline be built without further delay.

We are and we will continue to be a good customer, a good neighbor and a good trading partner with both Mexico and with Canada.

This nation will need to rely on a broad range of energy sources. The hard fact is that we depend on nuclear power now for 13 percent of all the electricity consumed in the United States. A few communities, for instance, Chicago, derive more than 50 percent of all their electricity from nuclear power plants. The recommendations of the Kemeny Commission investigating the Three Mile Island incident will help us to insure safety, but nuclear power must play an important role in the United States to insure our energy future.

In June I set an ambitious but important goal for meeting 20 percent of the nation's energy needs from the sun by the year 2000. With steeply rising OPEC prices and greater supply uncertainties, attainment of this goal is more important than it ever was.

No cartel can control the price of solar power. No country can embargo solar power. We've already tripled our federal investment in solar energy, and the new solar bank that I have proposed will permit all Americans to join in making widespread solar power use a reality.

The actions that I've already taken with Congress the last two years will reduce our projected imports of foreign oil by 4 million barrels per day. The new actions I've described for you last night and today will save us an additional 4.5 million barrels of oil—foreign oil—a day below what we are presently consuming by the year 1990.

I'm going to keep these initiatives moving, and every one of us—public official and private citizen—must keep up this pressure for progress. Our basic strategy is as clear as it can be. Together, you and I and every American are simply going to change the way this society creates and uses its energy, and as we do so we are going to find ourselves growing stronger, more free and more confident at home and around the world.

SYNFUEL POLITICS[6]

President Carter's call for a "fast track" development of synthetic fuels was, in large measure, a response to a political problem—his own waning prospects for reelection in 1980. Politics, therefore, is likely to determine just how far the "synfuels" program goes.

The ninety-sixth Congress has been notably timid and unproductive, given to darting off in a new direction with every shift of political current like a school of nervous fish. When the gasoline lines started forming early in the summer, Congress feverishly embraced synthetic fuels as an answer to take home to angry constituents. But when the lines disappeared in most of the country, the congressional will to deal boldly with the energy crisis began to dissolve into the irresolute bickering that has characterized the legislative branch over the past couple of years.

If the synthetic fuels program is to move forward at the

[6] Reprint of article entitled "The Current Politics of 'Synfuels,' " by Philip Shabecoff, reporter for the New York *Times*, Washington Bureau. *Technology Review*. 81:31. Ag. '79. Copyright © 1979 Alumni Association of the Massachusetts Institute of Technology. Reprinted by permission.

pace outlined by the President, it must be launched with the momentum of full national consensus. The technology to carry such a program to fruition exists, the policy makers believe. And there is little argument with the national security rationale for relieving dependence on foreign sources of energy.

Whether that consent is extended probably will depend on whether the public is willing to bear the economic, environmental and social costs involved, once those costs are made clear. It will also depend on whether, and how well, opponents of the massive synfuels program are able to make their case that there are less expensive, more efficient, and environmentally benign alternatives.

Economic issues are likely to form a particularly difficult political barrier to Mr. Carter's program. For one thing, the cost of developing the synthetics program is going to be high. According to an analysis by the Rand Corporation [a research institution, studying the areas of national security and public welfare], the crash program envisioned by the President could easily cost twice the $88 billion price tag he put on it. The capital absorbed by the program will drain funds that would otherwise be invested in different sectors of the economy. This could very well be a stimulant to continuing inflation as well as a dampener on improved industrial productivity.

Department of Energy officials insist that while there may not be solutions at the moment to all the environmental problems posed by toxics and carcinogens released during the extraction, processing and consumption of synthetic fuels, technology can provide the answers. More problematical, perhaps, would be resolving priorities in land use and distribution of water resources. On such issues, emotion as well as economics must be weighed in the political balance.

No one has suggested there is a solution at hand for the "greenhouse effect" of carbon dioxide released into the atmosphere during the combustion of hydrocarbons. The problem is often waved aside as something that can be dealt with in the distant future. But a recent study prepared for the Department of Energy suggested that CO_2 could start making a noticeable impact on world climate by the year 2035 (at the current rate of organic fuel consumption). With a massive shift to synthetic fuels, climatic changes could start as early as the end of this century, according to Dr. Gordon J. MacDonald, one of the authors of the report.

Environmentalists and others argue that the President's goal of reducing oil imports can be met by more intense conservation along with rapid development of solar energy and exploitation of the nation's large reserves of unconventional natural gas. They argue further that not only would scarce resources and public health be better protected, and costs be lower, but disruptive large-scale additions to the infrastructure, such as pipelines and new communities in rural areas, would not be necessary.

It will be some time before the intrinsic merits of competing energy strategies can be fully weighed. But a political decision will be made soon. The future of synthetic fuels development will probably be determined by how the issue is treated by candidates for office in 1980—and how the voters respond.

NEEDED: A BALANCED ENERGY PROGRAM[7]

Three questions must be answered at this point: What kind of program is needed? How can the program be financed? What would be the likely consequence? We answer each of these questions in turn.

Conservation

Of the two "new" energy sources, the most immediate priority is conservation, which can reduce America's dependence on imported oil until solar energy can make a substan-

[7] Excerpted from the conclusion of *Energy Future*, ed. by Robert Stobaugh, director of the Energy Project at the Harvard Business School, and Daniel Yergin, director of the International Energy Seminar at the Center for International Affairs. *Energy Future: Report of the Energy Project at the Harvard Business School.* Random House. '79. p 227–33. Copyright © 1979 by Robert Stobaugh and Daniel Yergin. Reprinted by permssion of Random House, Inc.

tial contribution. The conservation measures depend on the nature of the sector.

☐ *In transportation,* an obvious form of subsidy would be to experiment with free public transportation in some municipalities. But since the automobile is such a central part of American life, we doubt that even free public transportation would reduce gasoline consumption dramatically. The mandatory standards for fuel efficiency seem to be especially attractive because the government has to regulate and monitor only a few companies. Stronger standards, which do not hamper flexibility and experimentation, should be set for the post-1985 period. And a gasoline tax high enough to have a similar effect on consumption is politically unacceptable.

☐ *In the industrial sector,* there are many decision-makers, but they generally have better information than most private citizens about the real costs of alternative energy choices. To give them appropriate signals and incentives, we suggest offsetting payments in the form of investment tax credits and accelerated depreciation up to 40 percent of capital costs. This seems to be better than forcing conversion to coal because of the environmental and health problems caused by coal.

☐ *The residential-commercial sector* is highly fragmented. While the potential savings from retrofit of existing dwellings can be very great, market imperfections, in particular poor information and limited access to capital, currently present great barriers. Under these circumstances, we suggest tax credits of up to 50 percent of retrofit costs, with rebates for lower-income groups. Because it is desirable to use existing organizations where possible, electric utilities should be encouraged to deliver energy conservation. This, of course, means changing regulations to make the conditions attractive enough for them to do so. The new-building market can be reached through revised building codes and through the mortgage requirements of loan agencies.

Solar Energy

Solar energy presents different problems. Although the technology used in solar heating is widely known, the application of the technology has been very limited. It has long been recognized that the benefits to society of the diffusion of technology exceed the benefits that accrue to the entrepreneur. Individuals and companies are more risk-averse than society as a whole, and information generated by an installation that does not work is often of less use to the innovator who failed than it is to society. Therefore, solar energy deserves and needs greater governmental assistance than conservation. Thus, we suggest that at least until installations are commonplace, solar users receive an offsetting payment equal to about 60 percent of installation costs.

Such payments are especially justified for the smaller solar units for hot water and space heating that would be located on site since they can make an important contribution fairly quickly. It is unfortunate that Small Solar does not have the government support that high-technology programs like the power tower do. The high-technology programs have little potential to help the country in this century. The use of on-site solar installations in existing homes is hampered by some of the same barriers that slow residential conservation —homeowners who are short of capital and information, and averse to risk. Offsetting payments should be accompanied by an attempt to use existing institutions, such as utilities, both for delivery and for financing. For new homes, office buildings, and factories, a large subsidy also would likely prove effective. Again, as with conservation, a considerable long-term education campaign is required. For example, solar heating will certainly prove more cost effective if used in buildings in which a conservation effort has already been undertaken.

In order for conservation and solar incentives to work, certain additional barriers must be broken. At present, for instance, an industrial firm wanting to cogenerate electricity and steam can face a regulatory obstacle in selling electricity to its utility. Solar energy is hampered by a lack of standardized building codes, confusion over the right of a person to prevent others from blocking the sun, utility regulations that deny a solar house all-electric rates, and property taxes.

Our recommendations for conservation and solar energy are rough guides. It is impossible to determine precisely how large various subsidies should be, nor do we know what subsidies Congress and the Administration would approve. We do know, however, that they should be a good deal larger than any that have been approved or officially proposed so far.

Need for Government as Champion

If the nation is to make the transition to a more balanced energy system, the government must be the champion of conservation and solar for several reasons. First, the conventional energy sources have a host of allies, witting and unwitting, in those analysts who understate the external costs generated by these sources. They not only tend to underestimate the disadvantages of imported oil, but also tend to underestimate the environmental costs of conventional energy. Studies, for example, often equate the health costs of pollution with lost wages plus medical expenses. Presumably, in such a formulation it "costs" society very little for a nonworking wife to contract lung cancer from emissions if she dies quickly, so that large medical bills are avoided. Second, the conservation and solar energy industries do not have as many companies and workers involved in them as do other energy sources. Imported oil, for example, has a powerful constituency among those who produce, refine, and distribute it. The international oil majors take in more money in a few hours than the entire solar industry does in a year. And these companies, their workers, and their customers have a natural tendency to favor their ongoing activities.

The government must lead, for the only thing that is going to happen "automatically" in the years immediately ahead is an ever greater stream of imported oil. Some of the most efficient large enterprises in history manage that process in a way to make it seem quite easy, transporting oil halfway around the world and then refining it, all for just a few pennies per gallon. (Most of the price paid at the pump goes in the form of payments to the governments of the producing and consuming countries.) But government leadership does not mean government management. Rather, it means correcting market defects in a way to create more jobs and more business opportunities for both large established companies and small new firms with a stake in conservation and solar energy.

Are we not talking about a great deal of money? How to pay for this kind of program? Many oil and gas executives realize that a windfall tax on part of the profits that result from deregulation of old oil is likely. We would propose that the windfall tax be specifically assigned to financing—primarily by tax credits, but also by grants and loans—conservation and solar, which are the two most promising alternatives to oil. And as the windfall tax might be self-extinguishing as we move to a free market for domestic oil and gas, so these credits might be self-extinguishing over a period of, say, ten years—an important feature, because it has often proved difficult in the past to stop programs once they are launched. Our proposal would thus respond to two of the most urgent problems in US energy policy—the stimulation of conservation and solar, and the need to gradually free all oil prices, not just those of new oil.

As it is today, the system of price regulation is highly irrational. And if our analysis about the impact of larger, rather than smaller, oil imports is correct, then an irrational American pricing system could be one of the main causes of much higher oil prices in the years ahead, with all that will do to the Western economy. It makes no sense for the United States to be as integrated as it is into the world oil market—by far the largest consumer of OPEC oil—and yet have a pricing system that is partly insulated from the market.

Our proposal would respond to another need, a political need, if the nation is ever to resolve its energy difficulties. It would help to bridge the gap between contending parties in the bitter American energy debate. The various interest groups—oil and gas producers, advocates of conservation and solar, even environmentalists and consumerists—are secret allies, though, to understate the matter, not all by any means would recognize this truth. Oil and gas producers are convinced that they need higher prices in order to maintain production. Advocates of conservation and solar should recognize that they need higher prices to make their programs more attractive. Consumerists need somewhat higher prices now to help protect the public against the awesomely high prices that could eventuate if the United States does end up importing 14 million barrels a day. It is not merely rhetoric, but absolute necessity, to find some ways to make this alliance clear to the various participants. Our proposal reconciles their interests.

Moreover, we wish to stress the need for greater understanding among normally warring parties. For instance, public interest groups must understand the substantial and complex difficulties faced by utilities as they try to adapt to the new energy era. Utilities should be partners in the promotion of conservation and solar. The exclusion of utilities from the conservation business in the 1978 National Energy Act was, in this connection, not some minor mistake but a major blunder, creating a very significant and totally unnecessary barrier to the exploitation of conservation energy.

The Consequence of a Balanced Program

What pattern of energy use will result from the move to a more balanced system? Although the base of experience with incentives for conservation and solar energy is inadequate for detailed predictions, some studies have presented plausible estimates of fairly high energy savings—given propitious conditions—over a period of ten to twenty years. Even after generously discounting such estimates, the potential savings appear attractive. To illustrate, we compare a conventional forecast for the late 1980s with a *possible* pattern that could result from increased usage of conservation and solar energy. Even though the estimates for these two sources are far below some . . . , it is striking the extent to which conservation, with the aid of solar energy, rather than ever-increasing oil imports and domestic supplies of coal and nuclear, *could* help meet

US energy needs. . . . Conservation and solar energy would provide two thirds of the "increased" energy supplies, compared with only a quarter in the conventional program based on Department of Energy forecasts (that is, 11 million barrels daily in the balanced program, compared with only 4 million daily in the conventional program, out of a total new supply of 17 million). And imported oil in the balanced program would not increase beyond the 1977 level of 9 million barrels daily, whereas the conventional program foresees an increase of 5 million daily—to a new level of 14 million.

The balanced program would still mean considerable use of traditional energy sources. . . . But the balanced program would represent the beginning of a transition to alternatives that pose far fewer problems than increased reliance on imported oil.

Conventional domestic production does not offer the same opportunity. The matter can be viewed thusly: Our conventional energy production—oil, gas, coal, and nuclear—may be thought of as well-explored producing regions. We favor continuing and augmenting production in these terrains. But in terms of allocating resources and effort for further major increments of energy, the evidence strongly suggests that the nation would be better served by concentrating its exploratory and development "drilling" in the partially proven acreage of conservation, and the promising but still largely untested acreage of solar.

What is still missing is an energy policy to guide the transition. What we propose would make possible an economically sound and politically workable transition away from ever-growing dependence on imported oil. No other nation has so great an impact on the international energy system. Now is the time for the United States to come to terms with the realities of the energy problem, not with romanticisms, but with pragmatism and reason. And not out of altruism, but for pressing reasons of self-interest.

III.   THE NUCLEAR DILEMMA

EDITOR'S INTRODUCTION

With this section we turn from general policy recommendations to a consideration of several alternatives. Not many years ago, nuclear power was looked upon by many as the solution to the problem of generating electricity in an oil-short world. But the 1970s were a watershed for that view—and for nuclear power—as the pieces in this section explain.

During the 1970s, nuclear power made slow but measurable progress in the United States. By 1980, the nuclear power industry was producing 3 percent of the energy from all sources in this country. Nuclear power was by then accounting for 9 percent of all electricity Americans were using. Plans were on drawing boards in the United States and other western nations to double the output of nuclear power plants by the year 2000.

Events in 1979, however, put any increase at all in doubt. The most dramatic event was an accident that crippled Three Mile Island, a nuclear plant near Harrisburg, Pennsylvania, on March 28. By the year's end, the Federal Government had ordered US utilities to operate their sixty-seven nuclear reactors under stricter (and more expensive) safety measures. And the nuclear industry had to face the possibility that the ninety-two nuclear plants under construction would never be finished. Summing up the year's events, one nuclear engineer lamented: "It makes you wish 1979 never happened."

What went wrong? In this section's first article, Professor Carroll L. Wilson, former general manager of the Atomic Energy Commission (1947–1951), tries to answer that question. He concludes that no one was much interested in solving the problems of nuclear wastes. The rewards for engineers and nuclear scientists came from designing workable reactors, not from seeking a safe way to get rid of radioactive wastes.

The second selection, by Richard Munson, examines the cost of nuclear power. During the 1970s, the cost of building a nuclear power plant rose more than 25 percent a year. Munson, director of Solar Lobby, suggests that additional safety measures required of nuclear plants in the wake of the Three Mile Island accident will price conventional nuclear power out of the market.

Someday nuclear fusion—in theory a relatively safe way of producing usable energy—may give new life to the nuclear industry, and to the nation's hopes for generating electricity with nuclear fuel. But, as Isaac Asimov, noted author and biochemist, writes in this section's concluding article, that time—if it ever arrives—is at least a half century away.

WHAT WENT WRONG[1]

Soon after the civilian Atomic Energy Commission took over the Manhattan District on January 1, 1947, questions were raised about when atomic power, as it was then known, would become economic. Would it be when coal was $10 or $15 a ton? In the second year of the Atomic Energy Commission these inquiries became sufficiently insistent that we turned to the General Advisory Committee for an opinion. This committee, chaired by J. Robert Oppenheimer, included Enrico Fermi, J. B. Conant and others who were the most knowledgeable in the world at that time regarding atomic energy and its potential. They disavowed having any crystal ball but agreed to think about the question.

In 1948, they came back to us with the following estimate: if billions of dollars were invested in research and development; if industry entered in a large way; and if there were

[1] Article entitled "Nuclear Energy: What Went Wrong," by Carroll L. Wilson, Mitsui Professor in Problems of Contemporary Technology, emeritus, at Massachusetts Institute of Technology, and former general manager of the Atomic Energy Commission (1947–51). *Bulletin of the Atomic Scientists.* 35:13–17. Je. '79. Reprinted by permission of the *Bulletin of the Atomic Scientists,* a magazine of science and public affairs. Copyright © 1979 by the Educational Foundation for Nuclear Science, Chicago, Illinois.

quite a lot of luck; then, they said, twenty years hence, in 1968, half of the electric power plants being ordered might be nuclear.

This forecast is one of the most accurate I can remember. Billions were invested, industry entered in a large way, and there was a lot of luck, including the remarkable success of the submarine power plant, and in 1968—twenty years later—about half of the new electric power plants ordered in the United States were nuclear. Nevertheless, it wasn't until 1970 or 1971 that nuclear energy exceeded 1 percent of our primary energy supply and passed wood as a fuel.

What has happened to this unique source of energy? Between postponements and cancellations, new orders have been zero since 1975. A considerable part of this reduction reflects unexpectedly lower growth rates in electricity demand, leaving the electric power industry with excess capacity. However, lengthening delays in permits and licensing, changing safety and operating standards by governments, rapidly rising construction costs, quadrupling of uranium prices and lack of provision for spent fuel from existing reactors have been some of the factors which have caused nuclear plant orders to fall to almost zero. In addition, the increasing number of antinuclear objectors and the ambivalent attitude of the Carter Administration—"nuclear only as a last resort"—led to reappraisals by utility executives who must raise the $1 billion which each nuclear plant now costs.

Even though I have only been marginally involved in nuclear affairs for many years, recent events such as *The China Syndrome* film and the impact of the Three Mile Island nuclear power plant accident in Pennsylvania have led me to ask what has gone wrong? Nuclear power, potentially such a useful source of energy, is now beleaguered and rejected by broad sections of the public, not only in this country but overseas.

By 1948, nuclear reactors were already operating, including three very large reactors at Hanford, Washington, which were producing plutonium and through which a portion of the Columbia River was flowing and being slightly warmed.

It was said at the time that although the general parameters for a nuclear power reactor were known, we lacked engineering knowhow and knowledge of materials. In fact, no one spent much time then nor even much later looking at the total system of the nuclear fuel cycle from the fuel enrichment through the whole system of reactor operation, reprocessing of spent fuel, and disposal of radioactive wastes after recovering reusable products like uranium-235 and plutonium. All attention was focused on the reactor. There was no awareness that the whole system must function or none of it might be acceptable.

Apart from nuclear weapons, which were then being produced in small numbers, the most obvious military application was to submarines. Existing submarines had to operate on the surface most of the time using noisy diesel engines to charge batteries which allowed them to operate briefly when submerged. Nuclear fuel which might not need to be renewed more than once a year could provide continuous energy for submerged operations. The prospect of a submarine unlimited in range and speed for underwater operation held enormous potential. This was obvious even before submarines became the platforms for launching intercontinental ballistic missiles with nuclear warheads.

Thus the Navy placed a high priority on nuclear reactors from the beginning. Several naval officers went to Oak Ridge, Tennessee, in 1946 and 1947 to learn all they could about the possible kinds of reactors which would be highly concentrated in power density and able to fit into the very small space of a submarine. By 1948, several of the people who had gone to Oak Ridge were brought together in the Navy Department, Bureau of Ships, in a group headed by Captain Hyman Rickover. Frankly, power reactors were fairly well down on the list of our priorities in the Atomic Energy Commission. In those early days, we were involved with the troubles of producing fissionable material for weapons and in the overall problems of getting the young agency underway. My deputy, Carleton Shugg, pointed out that we would be under enormous pressure from the Navy to get on with the development of a suitable reactor for the submarine. At his suggestion we made an arrangement with Admiral Earle W. Mills, then chief of the Bureau of Ships, for Rickover to be given responsibility in the Atomic Energy Commission for the development of a reactor for a submarine (he was responsible for the submarine in the Navy). Rickover used his two hats, the Atomic Energy Commission and the Navy, to push the project with extraordinary diligence.

It was decided early to develop a pressurized water system involving a thick pressurized tank able to contain 2,000 pounds per square inch pressure and to use highly enriched uranium fuel in order to pack as much power density into the smallest space possible. A pressurized, light water design suggested by Farrington Daniels of the University of Wisconsin was selected as the basic concept.

In a remarkably short time this prototype reactor was tested in the Idaho desert. The prototype met the requirements and by the time the reactor was put into the submarine, it had been tested at full scale and the bugs worked out. The *Nautilus* was launched in 1954 and by 1958, several more had been launched including the *Sea Wolf*. Over the past quarter century nuclear submarines have built up a remarkable record for safety and reliability.

The point of reviewing this history is that the pressurized water reactor was peculiarly suitable and necessary for a submarine power plant where limitations of space and weight were extreme. So as interest in the civilian use of nuclear energy began to grow, it was natural to consider a system that had already been proven reliable in submarines. This was further encouraged by the fact that the Atomic Energy Commission provided funds to build the first civilian nuclear power plant at Shippingport, Pennsylvania, using essentially the same system as the submarine power plant. Thus it was that a pressurized light water system became the standard model for the world. Although other kinds of reactors were under development in different countries, there was a rapid scale-up of the pressurized water reactor and a variant called the boiling water reactor developed by General Electric.

These became the standard types for civilian power plants in the United States and were licensed to be built in France, Germany, Italy and elsewhere.

If one had started to design a civilian electric power plant without the constraints of weight and space as required by the submarine, quite different criteria would apply. For example, a shutdown of a month to refuel fits the annual overhaul of a submarine, whereas in a base load power plant one wants maximum operating hours per year and therefore continuous fueling is very desirable. Indeed such criteria were applied in the case of the Candu heavy water reactor developed by the Canadians and the Magnox gas-cooled reactor which became the backbone of the UK [United Kingdom] nuclear power program. Both were designed from the ground up as civilian nuclear power systems and both used natural uranium as a fuel. The Magnox system was regarded by most of the nuclear fraternity as big, ugly, primitive and inelegant. The Candu reactor in Canada had a large number of small pressurized tubes instead of one large pressure vessel and operated in a "lake" of heavy water.

The easiest route to commercial power was to take the submarine reactor, greatly expand its size with corresponding economies of scale. The acceptance of this approach was due in part to the vigorous development and sales promotion by Westinghouse and others among potential licensees in Europe and Japan. The comparatively weak promotion by the Canadians left Candu "in the dust." No one was attracted to Magnox.

Moreover, in each country the licensing and safety regulations were modeled after those in the United States to fit the pressurized water reactor and boiling water reactor systems. They did not fit Magnox or Candu systems regardless of their comparative merits and safety differences. By 1978, the installed capacity of Magnox in the United Kingdom was 7.7 gigawatts (electric) supplying 14 percent of UK electricity and the Candu capacity was 5.2 gigawatts (electric) supplying 22 percent of Canada's thermal generation of electricity. In the United States 50 gigawatts (electric) of light water reactors (pressurized and boiling water reactors) were operating, supplying 12 percent of US electricity.

I do not know the comparative behavior on failure of light water reactors versus Candu or Magnox systems which have different configurations and much lower power densities. Nor do I know how to obtain an objective comparison. Magnox and Candu plants have operated for many years without having episodes which hit the headlines.

From the very early days of the development of nuclear power, the attention was all on the front end of the fuel cycle and mostly on the reactor itself. Here's where the jobs and the career opportunities were. Here's where the commercial interest lay, not only in the reactors themselves but in all of the pumps and hardware and fuel assemblies that went with the reactor. These were projects of primary interest to physicists and engineers and billions of dollars were invested with hopes of profitable returns.

Because fast breeder reactors require the chemical reprocessing of the spent fuel in order to get plutonium, the reprocessing of spent fuel from pressurized water reactors and boiling water reactors belatedly received attention. Some years ago, the Atomic Energy Commission tried to encourage commercial interest in the business. Several firms, one at West Valley, New York, and another, a consortium at Barnwell, South Carolina, invested large sums of private money in building reprocessing plants. General Electric is said to have spent $50 million before abandoning their project. However, the chemistry and chemical engineering had received very low priority over the years and the processes were really not suitable for dealing with oxide fuel. Existing processes had been applied in a weapons program for dealing with metal fuel elements. Nobody had been much interested in the back end of the fuel cycle. The part that probably had the lowest priority had to do with the disposal of high-level radioactive waste from the chemical reprocessing operation.

Even in my day in the Atomic Energy Commission (1946–50), we spent $250,000 a month sinking steel tanks in the desert in Richland, Washington, to hold the high level waste which remained after the plutonium and uranium-235 had been extracted from the fuel elements of the Hanford reactors. Now thirty years later these tanks are very old. I wonder what their halflife is! [Halflife is the time required for one half the atoms of a given amount of radioactive substance to disintegrate.]

Chemists and chemical engineers were not interested in dealing with waste. It was not glamorous; there were no careers; it was messy; nobody got brownie points for caring about nuclear waste. The Atomic Energy Commission neglected the problem. Very little solid work was done on the conversion of high-level waste into solid nonleachable form nor on how to store it. It is true that a program of studies led to selecting a Kansas salt mine as a storage place but it turned out that it was not suitable. The central point is that there was no real interest or profit in dealing with the back end of the fuel cycle.

It is a little surprising that no group seemed to have looked at the total fuel cycle as an interdependent system. No one appeared with enough clout to set priorities and to argue persuasively that if some critical part of the system were missing perhaps the whole system would come to a grinding halt. Or again, substantially improving uranium usage was not given a high priority because the fast breeder reactor was always expected to come along and multiply the fuel supply so abundantly that getting higher burn-ups and more energy out of a given pass of fuel through the reactor was not considered an urgent problem.

Another source of uncertainty was the rapid scale-up of nuclear power plants before there was much operating experience with the small ones. The *Nautilus* reactor was perhaps 60 megawatts electric and Shippingport was not much bigger. The next power plant, Yankee (in Massachusetts), was about 220 megawatts electric—a three- or four-fold increase over the submarine power plant. The next jumps were to 600 or 900 or 1,000 megawatts. Essentially, this represented a scale-up of the same system with everything just thicker and bigger and more powerful including the very large pressure tanks with very thick steel walls to contain the 2,000-plus pounds per square inch and enough margin for surges.

The enormous heat transfer rates needed in light water reactors meant that uninterrupted coolant flow was absolutely indispensable. Furthermore, such flow had to be at very high velocity because if not, the fuel elements would heat up quickly and begin melting. Thus one of the nightmares that goes with the light water reactor is the loss of cooling accident (LOCA) which has been much talked about and studied on paper but never tested experimentally. It was such studies of these phenomena that led to coining the term, *China syndrome.* It was estimated that if the coolant circulation really failed through massive breaks of primary coolant pipes or pumps and failure of the various back-up devices, the core would begin to melt. As it melted, it would remain extremely hot and continue to melt right down through the entire structure and into the earth, releasing very large amounts of radioactive products into the atmosphere. Such a loss of cooling accident was called the "China syndrome" because if it went far enough, the core might go right through the earth (at least start that way). A LOCA accident could cost the lives of thousands of people and drop radioactive debris over thousands of square miles downwind of the accident according to an AEC study of the early 1960s. One of the grievous errors of the Atomic Energy Commission in the 1960s was the failure to carry out the repeatedly recommended experiment of running a reactor to destruction (LOCA) to find out what would really happen instead of depending on studies based on computer models for such vital information.

In adopting the submarine power plant as the chosen design for civilian nuclear power stations the industry failed to adopt the compact man-machine control system of the submarine. Instead they carried over the control room arrangement and philosophy of a conventional power plant. This involves a forty-foot long array of instruments and control handles which takes a crew of six to monitor and manipulate. In the submarine the whole system—reactor and boat operating in three dimensions—is controlled by the skipper overseeing several operators all in a space 10 by 12 feet. The philosophy and man-machine coupling resembles the system in a Boeing-747. All the essential gauges, switches, indicator lights and controls are readily visible to and controllable by one very highly-trained and highly-paid person in the case of the B-747. The submarine and the B-747 are much more complex systems than a nuclear power plant. Why aren't there integrated, one-man controls for a nuclear power plant, especially for the critical start-up and shut-down of the reactor?

It seems to me that there has been a failure to recognize that a nuclear reactor is a wholly different "kind of animal" from fossil-fueled boilers in which the fire goes out and the heat stops when you stop feeding gas, oil or coal. On the contrary, the heat does not stop in a nuclear reactor when the control rods are put in—the reactor still produces 10 percent as much heat as at full power and this dies away gradually.

Thus, quite aside from its much higher unit heat transfer rates a nuclear reactor is a very different system especially at "take off" and "landing." These differences can be summed up by the observation that one cannot imagine an accident in a fossil fueled power plant which would endanger anyone beyond the plant boundaries. On the contrary, after Three Mile Island the whole world is aware of dangers to which thousands are exposed by a nuclear plant shutdown or "landing" when there is a combination of operator errors, malfunctions of safety devices and wholly unforeseen phenomena such as the hydrogen bubble. How could all this happen?

Six years ago in an article for *Foreign Affairs*, "A Plan for Energy Independence," I wrote that I could not see how the issue of nuclear reactor safety was going to be resolved.

Like others who have followed closely the development of nuclear fission as an energy source for more than twenty-five years now, I originally and for some time believed that it could, without undue difficulty, become the most important source of energy we have, especially for electricity. But problems have mounted, and delays, restrictions and technical uncertainties have dogged nearly every one of the many steps needed to bring a new nuclear plant into full operation, thus drastically slowing down the nuclear input to our energy system. The determined opposition of states and localities and citizen action groups, plus rising caution by the Atomic Energy Commission, has stretched out to ten years the interval between application for a plant permit and bringing the plant "on line" at an economic power level.

In part, the political forces at work reflect an exaggeration of the problems, or at least a failure to weigh fully the inevitable trade-offs between energy supply and other factors. But these politically reflected concerns do have a substantial basis, both as to safety and as to unnecessary and unacceptable environmental consequences. Only if we deal with these factors can nuclear fission play the role I believe it must play in our total energy picture by 1985.

On safety, a real uncertainty now exists concerning possible accidents which could have disastrous consequences—especially the failure of liquid cooling systems resulting in a meltdown of the highly radioactive core and release of the gaseous fraction of these radioactive products into the atmosphere. A year of hearings by the Atomic Energy Commission has not persuaded the critics that current reactor plants are safe against such accidents, and the problem exists as well in the liquid-metal-cooled breeder reactor designs.

As I see it, the only way to meet these objections and so resolve the current impasse is to put all new plants underground. This is an entirely practicable course of action. Studies indicate that placing nuclear power plants underground would add only a small fraction to their cost. The extensive know-how of the mining industry plus that of the underground gas storage industry could be applied in placing such plants in suitable geological formations five hundred or more feet underground near load centers. So located, with suitable locks in the elevator shafts to contain and hold back any pressure of radioactive gas in the event of an accident or a meltdown, these plants could meet the requirements for nuclear safety. Placed in a suitable impermeable geological formation, a meltdown, even if it buried itself below the underground chamber level, would not leak radioactive products into underground water or into the atmosphere.

The second big nuclear production problem today is primarily environmental; it concerns the effects of the water discharge from large reactors in heating up streams and larger bodies of water, thus altering the ecology in many harmful ways. Here Europe has pointed the way to the answer—large cooling towers, built on the surface to recycle and cool the hot water discharges with very low net water heating.

Finally, there are problems in the safety and security of handling and transporting plutonium, and in the perpetual storage of radioactive wastes. In these areas, risks cannot be totally eliminated, but they can and should be sharply reduced to an acceptable level by determined action—as we enter an era of massive production, transport and handling of plutonium, which is one of the most toxic substances known.

All in all, the measures required to permit expansion of nuclear-fission plants will not be cheap or easy. But if the necessary steps are taken, nuclear-fission plants should be able to provide roughly 10 percent of our total 1985 energy needs at tolerable levels of risk and bearable costs. And experience in this next decade should tell us much about the degree to which we can hope to expand, by the end of the century, our use of nuclear fission, especially through the breeder reactor; in the 1985 time frame, the breeder is not likely to make a significant contribution.

Shortly after the article was published, I was in Stockholm and later in Japan. In both countries, nuclear programs were stalled because of public opposition to the reactor sites. I proposed to officials in both countries the alternative of putting nuclear reactors underground. Suitable underground sites would give positive assurance that the radioactive products of a loss-of-cooling accident would not be released to the atmosphere and endanger and contaminate areas downwind of the plant. There was absolutely zero interest on the part of anybody anywhere in doing so. The worry of governments and industry was that to put future reactors underground would imply that reactors currently operating were unsafe. Even when I pointed out that hundreds more reactors would exist in the future, no one was impressed. I also pointed out that the decommissioning costs would be greatly reduced. No sale!

To this observer, the most striking fact in this retrospective look at our nuclear energy program has been the lack of awareness that the whole interdependent system must be satisfactory. Such a view would have led to incentives for working on the back end of the fuel cycle. No one appeared to understand that if the whole system did not all hang together coherently none of it might be acceptable. Indeed, this is turning out to be the case. We seem to be years away from a real solution to the back-end problem. There are laws in three states of the United States and court actions in Germany which provide that until the back end of the fuel cycle including the disposal of wastes is satisfactorily assured no licenses will be issued for new reactors.

The interdependence of the total system is clear now but why wasn't it long, long ago?

NUCLEAR'S COST: TOO HIGH?[2]

Twenty-five years ago Atomic Energy Commission chairman Lewis Strauss predicted that nuclear power would be "too cheap to measure." Although admitting that Strauss may have been a bit optimistic, the nuclear industry continues to spend millions of dollars each year to convince consumers that nuclear energy saves them money. There is growing evidence that it does not.

Since 1974, utility executives have canceled orders for 32 reactors and deferred orders for 150 more. And some prominent Wall Street firms have advised their investors to beware of utilities planning nuclear power plants.

At the same time, Government forecasts of the nuclear potential have dropped dramatically. In the early 1970s, officials estimated that the United States would have 1,500 large reactors by the year 2000. Only a few years later, the projection was reduced to 400, and unofficial estimates now place the likely number below 300. (At present there are 70 licensed reactors; 89 more are in construction and 39 under review by the Nuclear Regulatory Commission.) The nuclear construction industry is in such a recession that, according to a trade journal, a new order is as rare as "the appearance of a comet." The recent near-disaster at the Three Mile Island nu-

[2] Reprint of article entitled "Nuclear Power: The Price Is Too High," by Richard Munson, director of Solar Lobby, a group working to promote the use of solar energy. *Nation.* 228:521+. My. 12, '79. Reprinted by permission. Copyright © 1979 The Nation Magazine, The Nation Associates, Inc.

clear facility in Pennsylvania has merely compounded the industry's financial woes.

Nuclear power has long been controversial on environmental and safety grounds. But as utility bills continue to climb, the controversy has become political. Meldrim Thomson, a strong advocate of reactors, was defeated last November when he ran for reelection as governor of New Hampshire. Thomson's public support had dropped significantly after he approved a 17 percent utility rate hike in order to help finance the embattled Seabrook nuclear power plant.

Since 1957, when the first United States reactor began producing electricity at Shippingport, Pennsylvania, the economics of nuclear power have been mixed. One of the successful ventures, Northeast Utilities' Connecticut Yankee plant, cost only $189 per kilowatt of capacity and has operated at 73 percent of capacity for more than nine years. At the time, a standard coal plant similar in size to the Connecticut Yankee project produced energy at $150 per kilowatt.

There are, however, frightening cases of economic failure. For example, the construction cost of the Pilgrim nuclear plant near Plymouth Rock increased from $65 million in 1966 to $239 million in 1972—nearly four times the original estimate. These cost overruns resulted in upping the per kilowatt price from $98 to $360.

But comparing current successes and failures is not the crux of the nuclear debate; the real issue is whether nuclear power will become economical in the future. And, looking ahead, it is increasingly difficult to predict that it will. Since the Connecticut Yankee went on line in 1969, capital costs have increased six times, the price of the fuel has more than quadrupled, fuel enrichment costs have skyrocketed and, inasmuch as no satisfactory means of waste disposal has yet been found, that cost is still unknown.

To arrive at a true balance sheet for nuclear power, the hidden costs, those subsidized by the Government, must be included. Government subsidies have been, and remain, essential to the nuclear power industry. In 1952, David

Lilienthal, the first chairman of the Atomic Energy Commission [AEC], outlined the risks associated with nuclear power and stated: "I wouldn't advise anybody who is responsible for private investment under present conditions to put his money or the money of those who rely on him into the development of power plants employing nuclear fission as a source of heat." The infant industry needed a cushion, and the Government obliged with a series of acts designed to encourage nuclear power development.

First, Congress approved the Price-Anderson legislation of 1957, relieving the nuclear industry of any liability problems arising from the risks of catastrophe associated with nuclear power generation. Second, the Government began pumping billions of dollars into nuclear research and development. Third, the Government subsidized, or permitted the industry to ignore, the costs of constructing fuel enrichment facilities, of regulation, of waste disposal and health costs associated with increased environmental radiation. A recent study by the Battelle Institute [nonprofit, public-purpose scientific research organization] for the Department of Energy conservatively estimates that the industry has received at least $18 billion in government subsidies. If this sum had been allocated equally over the cumulative output of nuclear plants since 1956, the cost per kilowatt-hour of nuclear electricity would have risen approximately 2 cents, a 95 percent increase. And even this estimate ignores the expenses of the Price-Anderson insurance legislation and of federal uranium policies.

Disregarding the subsidies, nuclear proponents tend to focus only on the capital needed to build and fuel a nuclear reactor. But over the past few years, these costs alone have zoomed to the point that Donald Cook, chairman of American Electric Power, the largest utility in the United States, commented, "an erroneous conception of the economics of nuclear power" sent utilities "down the wrong road. The economics never materialized—and never will materialize."

Nuclear construction costs are increasing so rapidly that accurate estimates are difficult to make. In 1967, the AEC predicted that reactors would cost $134 per kilowatt of generating capacity. By 1976, construction expenses increased to as much as $645 per kilowatt and current estimates are approximately $1,000 per kilowatt. One official of the Atomic Industrial Forum, a trade association, noted, "estimating capital costs for power plants is like shooting at a moving target."

According to Richard Morgan of the Environmental Action Foundation [an environmental research organization],

the dramatic increase in nuclear construction costs cannot be explained by inflation; since 1964, nuclear costs have increased more than ten times faster than the Consumer Price Index. Nor can rising costs in the construction industry be responsible; the cost of building an oil refinery has increased only one-tenth as fast as the cost of a nuclear reactor, and coal-fired plants have increased less than half as fast.

The only satisfactory explanation for these skyrocketing costs is that nuclear reactors are dangerous and complex. The utilities have often caused themselves expensive delays by not complying with federal safety regulations. In 1973, for example, twenty-nine of the thirty completed reactors were held up because of unsafe equipment or plant design. Physicist Amory Lovins estimates that nuclear power requires a total investment of $3,000 per kilowatt of net, usable delivered electric power. In other words, the power for a single 100-watt bulb requires a $300 investment. Projected nuclear growth in the United States through the year 2000 could absorb more than one fourth of the nation's entire net capital investment. Such extensive use of scarce capital is coming under increased attack.

Utilities also tend to ignore the costs on the opposite side of the construction process: the decommissioning of a nuclear plant. Some executives admit that dismantling a large reactor could cost $100 million; but even that may be optimistic. Elk River, a small experimental reactor in Minnesota, was dismantled for $6.9 million—almost $1 million more than it cost to build the plant. (Since Elk River was dismantled after only a few years, inflation was not a major factor.) The costs of decommissioning the contaminated plant at Three Mile Island over the next four or five years are as yet incalculable.

Not only are nuclear plants expensive to build and raze; they also produce far less electricity than was anticipated. On average, nuclear plants have delivered less than three fourths of the electricity they were designed to produce. The power stations cannot run constantly at full capacity; they must close for at least a few weeks each year for maintenance, refueling or repairs. In addition, some plants are required to operate at reduced capacity for safety or environmental reasons.

Reactor manufacturers and the AEC had promised utilities that nuclear plants would operate at 80 percent of capacity. A 1976 study by the Council on Economic Priorities found that on the average the plants have run at only 59 percent of capacity. In other words, the nation's nuclear plants have been out of service more than 40 percent of the time, about twice the outage rate planned for by the utilities. By contrast, coal-fired plants ran at 75 percent of capacity.

The champions of nuclear energy prefer to slide over high capital costs and low performance, and to emphasize instead the low fuel costs. And, indeed, in 1973 a pound of refined uranium oxide, or "yellowcake," which sold for $7, could produce as much electricity as 2.5 tons of coal costing about $20 a ton.

But by 1978, the cost of a pound of yellowcake had reached $52. Despite considerable prospecting, no significant new uranium deposit has been discovered in the United States since 1965. Therefore, since mining companies must extract lower quality ores, the cost of yellowcake will continue to rise. At approximately $100 per pound, uranium fuel loses its once touted advantage over coal or oil.

There are, of course, overseas sources, but these are creating some of the conditions that now plague the petroleum market. In 1972, several uranium mining firms met secretly in Johannesburg, South Africa, to "discuss ways and means of assuring an adequate price for uranium." The group, dubbed "UPEC" by nuclear critics, held subsequent meetings in Paris, Toronto and Chicago.

The cost of the nuclear fuel cycle does not stop at the mines. For use in a reactor, the active fuel within yellowcake must be enriched from 1 to 3 percent. Enrichment, an energy-intensive and expensive technology, has more than doubled in price since 1973. The Government's three enrichment plants cost billions of dollars to build in the 1940s and 1950s and use more than 2 percent of the nation's total consumption of electricity.

Thus, to conserve uranium supplies and avoid expensive enrichment, nuclear proponents have pushed for recycling or reprocessing projects. In theory, reprocessing would recover 25 percent of a reactor's spent fuel. Unfortunately, the process has not worked, and the attempts to make it work are getting more expensive. In 1976, Nuclear Fuel Services abandoned its West Valley, New York plant after learning that necessary alterations would cost approximately $600 million (twenty times the plant's original cost). In the same year, General Electric abandoned its Morris, Illinois reprocessing plant rather than spend the additional $100 million needed to make it work. The only current attempt to set up a commercial reprocessing plant is being made by Allied General Nuclear Services at Barnwell, South Carolina. It was to have been completed in 1974 at a cost of $100 million, but the estimate now is approximately $750 million. The *Wall Street Journal* labeled the Barnwell plant "one of the biggest white elephants of the nuclear age."

Along with reprocessing, the nuclear industry faces the technical and economic problems of radioactive wastes. To date, no acceptable method of long-term disposal has been devised and temporary Government storage facilities are inadequate. According to the Environmental Protection Agency, "even assuming the technical capability exists to insure total containment for the hazardous lifetime of the nuclear wastes, the cost of implementing the means to contain these wastes is enormous." As an example, the Department of Energy estimates that the price for safe disposal of 500,000 gallons of highly radioactive waste could come to more than half a billion dollars.

Not surprisingly, the rapidly inflated price of nuclear power construction, operation and fuel cycle is being translated into higher utility rates. The Environmental Action Foundation and the Critical Mass Energy Project [a consumer group] have analyzed the rates of the 100 largest electric companies between 1973 and 1977 and have found that those firms with 5 percent of more of nuclear generating capacity increased their rates 27 percent more than did firms without reactors.

But despite the poor economic outlook, some utilities and manufacturers are strongly urging further nuclear development. This peculiar phenomenon can be explained by the peculiar nature of power company operations. As regulated monopolies, electric utilities are not subject to the normal pressures of competition. State regulatory commissions attempt to set rates that are designed to cover a utility's operating costs *plus* a profit on its investment comparable to the profits earned by competitive business with similar risks. The size of the profit depends on the size of the utility's investment; the larger the investment, the larger its profits. Thus, utilities have a built-in motive to overbuild their plants and to adopt expensive (capital-intensive) technologies like nuclear power.

Current regulatory measures also make it easier for utilities to finance expensive investment in nuclear energy. Until recently, investor-owned utilities, like all profit-making businesses, had to borrow money to finance new equipment. Some commissions, however, are allowing utilities to increase their rates in order to gain the necessary capital. A procedure called "construction work in progress" or CWIP allows utilities to include power facilities that are under construction in their rate bases. The result is an approximate 15 percent increase in electric rates and a further boost for utility profits.

In addition, federal tax laws encourage a utility to go nuclear. Like other private businesses, utilities annually receive billions of dollars in tax breaks for investing in new equipment. But an obscure federal tax code allows power companies to keep their tax savings rather than pass them on to consumers. The Environmental Action Foundation found that in 1976 utilities collected from their customers $2.1 billion in "phantom taxes" that they never paid to the Government. The more money a utility invests in construction, the more phantom taxes it can collect. And nuclear power plants are the most expensive investment a utility can make.

Fortunately, the alternatives to nuclear power are beginning to gain increased attention, even from economists. According to Denis Hayes of the Worldwatch Institute, "dollar for dollar, investments to increase energy efficiency save more than expenditures on new power plants will produce." Hayes believes that, without decreasing the standard of living, the United States can decrease its total energy consumption almost 50 percent by eliminating energy waste.

Current economic studies of price increases also show that coal-fired plants are competitive with nuclear reactors. In 1976, for example, the Congressional Research Service predicted that then-planned coal-fired and nuclear plants would produce electricity at approximately the same cost. After a thorough economic review in 1976, Harvard economist Irvin Bupp stated, "The only way you can conclude nuclear power will be cheaper eight to ten years from now is to make systematically optimistic assumptions about nuclear costs and be systematically pessimistic about coal." And after evaluating the operating records of existing nuclear plants, the Council on Economic Priorities [which researches and disseminates information on social responsibilities of corporations, especially in such areas as environmental impact and energy planning] predicted that, for most of the nation, new coal-fired plants would produce cheaper electricity than new reactors.

Solar energy is also becoming more attractive. Unlike fossil and nuclear fuels, sunlight cannot be exhausted, and since it can be used in decentralized facilities, the high costs of shipping conventional fuels and transmitting power could be eliminated (transmission and distribution today account for approximately 70 percent of the cost of providing electricity to the average residence in the United States).

In January 1978, the Department of Energy found that solar water and space heating was competitive with nuclear electricity and fuel oil in the four cities studied: Boston, Washington, Grand Junction, Colorado, and Los Angeles. With the approval of solar tax credits, the study indicates that solar power is likely to be marginally competitive with natural gas.

The most exciting solar application is the photovoltaic cell that converts sunlight directly into electricity. Designed for the space program, the first solar cells were expensive— approximately $200 per peak watt in the late 1950s. Further research has brought the price down to $6 per peak watt. If a demand were created great enough to permit mass production, researchers believe the cost would drop even more dramatically. A recent United Nations report, for example, concludes that solar cells will become cheaper than nuclear power (approximately $1 per peak watt) if they receive a total investment of $1 billion—less than the cost of just one large nuclear plant.

The potential from other renewable sources is also significant. The Army Corps of Engineers recently estimated that adding turbines to existing small-scale dams would economically generate as much electricity as the United States currently receives from nuclear power. And as businesses and small entrepreneurs turn their attention to wind and biomass technologies, the price of solar energy will continue to drop.

According to a recent federal task force, solar sources could produce up to 30 percent of total United States energy by the year 2000. This level is more than is now contributed by natural gas, more than by coal and nuclear power combined, and more than by our imported oil. [For more information on solar energy, see Section VI, below.]

Citizens' groups around the country are beginning to master the complexities of energy economics. In addition to complaining about reactor safety, ratepayers are demanding that their state utility commissions investigate the wisdom of nuclear investments. The authority that utility commissions exercise over nuclear power varies according to state laws. Some of them control utility financing and could refuse the company permission to issue stocks and bonds for nuclear construction. Most commissions must also issue licenses before a utility can build any major new power facility. Before granting a license, the commissions must evaluate whether the plant is the most economical way to generate the needed electricity. In July 1978, the Wisconsin Public Service Commission, citing problems with uranium availability, waste disposal and decommissioning, ordered the utilities in the state to include no new reactors in their future plans. Also in 1978, the California Energy Commission canceled plans for the Sun Desert nuclear reactor, noting that the needed energy could be more cheaply "produced" through conservation and alternative sources.

Many consumer and environmental groups have effectively used economic arguments against nuclear power during hearings on a utility's proposed rate increase. In response to "unfavorable" commission decisions on rate hikes, Florida Power and Light, Detroit Edison, and Consumers Power in Jackson, Michigan canceled their nuclear projects. In 1976, the Wisconsin Public Service Commission granted the Madison Gas and Electric Company an $8 million rate increase only after the utility had agreed to drop plans for the Koshkonong nuclear plant. Also in 1976, the Seattle City Council overruled the municipal utility's plans for two nuclear plants, opting instead for conservation. One council member stated, "Conservation is the lowest-cost method of generating power, and can be accomplished through sensible management without affecting our standard of living." [For further information on conservation, see Section VII, below.]

Some citizens' groups have also successfully convinced their commissions to shift the financial risks associated with nuclear energy from consumers to stockholders. In 1976, for example, Safe Power for Maine gained a refund of $3 million in replacement power costs collected by Central Maine Power during a forced outage of the nuclear plant. The commission reasoned that the utility should collect damages from the manufacturer of the faulty equipment that caused the outage.

Despite its many environmental and social disadvantages,

nuclear power has received public support because it was
thought to be cheap. But a recent Harris poll shows that a 2-
to-1 majority would oppose reactors if they believed nuclear
power was more expensive than other sources. Clearly, more
and more Americans will oppose nuclear power as cost increases and government subsidies are added to their utility
bills. The nuclear industry, already on shaky economic
grounds, may not survive the challenge.

NUCLEAR FUSION: THE PERFECT SOLUTION?[3]

The world faces a crisis that may destroy civilization in
our own lifetime. It is usually referred to as an energy crisis,
but it isn't.

It is an oil crisis. The earth's oil wells may begin to run dry
in thirty years, and without oil it would seem that the industrial world will clank to a grinding halt and that there will be
no way in which the teeming population of the world could
be supported.

Yet who says oil is the only source of energy? It is, at present, the most convenient source; at present, the most versatile. Matters do not, however, have to stay in the present.

The early decades of the twenty-first century may see oil
supplies at a useless trickle and yet find energy plentiful and
electricity coursing through the nerves and veins of industry.
With plentiful, unending electricity, we could even manufacture our own oil to fill indispensable needs: Electricity can
break down water to hydrogen and oxygen; the oxygen can be
discarded, and the hydrogen can be combined with carbon
dioxide from the air to form gasoline. The gasoline can then
be burned and will combine with the discarded oxygen to
form water and carbon dioxide again.

[3] Reprint of article entitled "Nuclear Fusion: Where to Get Energy When the
Oil Wells Run Dry," by Isaac Asimov, biochemist and science writer. *Parade.* p 4–5. F.
18, '79. Copyright © 1979, Parade Publications, Inc. Reprinted by permission of
Parade Publications, Inc. and of the author.

Nothing will be used up but electricity, and the electricity can come from the greatest and most copious source of energy on our planet—the hydrogen in seawater.

That hydrogen represents the great ark in which humanity can ride out the oil shortage that now threatens to overwhelm us and come to rest finally on the quiet uplands of energy-plenty.

There is a catch. The ark is not yet quite within reach. Our hands still grope for it, but we cannot yet squeeze the energy out of hydrogen.

The simplest way of getting energy out of hydrogen is to combine it with oxygen—to let it burn and deliver heat. Such a process, however, involves merely the outermost fringe of the hydrogen atom and delivers only a tiny fraction of the energy store available at its compact "nucleus."

Something other than hydrogen-burning—something much more dramatic—takes place at the center of the sun. Under enormous gravitational pressures, the substance at the sun's core is squeezed together, raising the temperature there to a colossal 15 million degrees Centigrade (24 million degrees Fahrenheit).

At such pressures and temperatures, the very atoms of matter smash to pieces. Their outer shells break away and expose the tiny nuclei at the center, which then drive into each other at thousands of miles per second and sometimes stick. When hydrogen nuclei stick together to form the slightly larger nuclei of helium atoms, the process is called "hydrogen fusion."

Every second, 650 million tons of hydrogen are fusing into 654.4 million tons of helium at the sun's center. This process produces energy. Each missing 4.6 million tons per second represents the energy that pours out of the sun in all directions. A very small fraction is intercepted by the earth, and on that energy all life is supported.

Though it takes an incredible amount of hydrogen fusion each second to support the sun, there is so much hydrogen in that giant object that, even after some five billion years, it is still mostly hydrogen. The sun can continue to produce energy for perhaps seven billion more years before its fusion mechanism begins to falter.

Can we somehow take advantage of this process on earth?

The trouble is we can't duplicate the conditions at the center of the sun in the proper way. To begin with, we need enormous temperatures.

One way of achieving such temperatures is to explode an "atomic bomb" that is powered by uranium fission. For just a brief period of time, temperatures in the millions of degrees are produced at the center of that explosion. If hydrogen in some appropriate form is present there, it will fuse. The result is that the atomic bomb becomes the trigger for the greater blast of a "hydrogen bomb."

The Need for Controlled Fusion

Naturally, we can't run the world by exploding hydrogen bombs. We want *controlled* fusion—the kind that produces energy a little bit at a time in usable, nondestructive quantities.

One way would be to start with a small quantity of hydrogen and heat it until it fuses. There would only be a small amount of energy produced. This could be bled away while new hydrogen is added to the mix to undergo fusion in its turn.

Heating hydrogen to the required temperature isn't easy, but it can be done by electric currents or by pumping in energetic subatomic particles. The trouble is that hydrogen expands as it's heated, and its atoms drift irretrievably away in all directions. We must hold the hydrogen in place while it is being heated. But how? The sun holds its hydrogen in place with its enormous gravitational field, but we can't imitate the sun's gravity on earth.

Nor can we force the hydrogen to remain in place by keeping it in a container. The heat might cause the container to vaporize. On the other hand, if we kept the container cool while the hydrogen heated up, the hydrogen would lose heat again upon contact with the cool container.

One possibility is to use a magnetic field. A magnetic field is not matter and is neither hot nor cold. As the hydrogen is heated, its atoms break down to electrically charged fragments, and these are repelled by the magnetic field. The fragments can't break through the magnetic field and must stay in place.

The problem is designing a magnetic field of the proper shape and intensity that will remain stable and not spring a leak. It's not an easy job. Scientists in the United States, Great Britain and the Soviet Union have been working at it for nearly thirty years. The best device proposed thus far is the "tokamak," first developed in the Soviet Union.

But even a tokamak won't do the job for ordinary hydrogen. In the center of the sun, a temperature of 15 million degrees Centigrade is sufficient because the hydrogen is squeezed together very densely. On earth we must work with much thinner gas, and that requires still higher temperatures.

Fortunately, there is a kind of hydrogen easier to fuse called deuterium. Only one out of every 6,500 hydrogen atoms is deuterium, but even so there is enough in each gallon of seawater to equal the energy supplied by burning three hundred gallons of gasoline. Since there are 3.6 quintillion gallons of seawater on earth, there is enough deuterium to last billions of years at the present rate of energy use.

The temperature required can be lowered further if a still-rarer kind of hydrogen called tritium is added to the deuterium. Tritium is radioactive and hardly occurs in nature, but it can be manufactured in the laboratory.

If a quantity of deuterium-tritium mixture is made dense enough, heated hot enough and kept in confinement long enough, it will fuse. There are well-worked-out figures for what is needed for all three conditions, and scientists have been edging toward the critical combination. Recent work with tokamaks at Princeton University and the Massachusetts Institute of Technology has confirmed that fusion induced by magnetic confinement is a real possibility—once a better tokamak can be built. But this, experts say, is still many years away.

But magnetic confinement isn't the only route to fusion

power. It's only needed when the hydrogen is heated slowly and would therefore expand and drift away while it is being heated.

Suppose the hydrogen were heated very rapidly. It might then reach fusion temperature so rapidly that the hydrogen has not time to expand before it starts fusing. That's what happens in a hydrogen bomb. The uranium fission develops its high temperature so rapidly that any hydrogen present fuses before it can scatter.

We can't use a fission bomb for controlled fusion, however. Some other way must be found to raise the temperature very rapidly.

One way is to make use of a laser. Lasers, first developed in 1960, produce light in a very tight beam. The total energy may not be unusually great, but the beam can be focused on such a microscopic point that the concentrated energy raises the temperature at the point to millions of degrees in a fraction of a second.

Imagine a mixture of deuterium and tritium inside a tiny, thin-walled glass bubble. If the bubble is struck simultaneously by a number of laser beams from different directions, the heating takes place all around the outer skin of the bubble. What expansion there is forces the gas upward. The inner portion of the bubble goes way up in density, further up in temperature, and begins to fuse.

We can imagine bubble after bubble dropping into position and being fused by accurately timed bursts of laser light. Work is underway at the University of California's Lawrence Livermore Laboratory to determine the feasibility of laser fusion.

Of course, it takes considerable energy to keep the lasers going, and they are expensive devices. Simpler and more efficient might be beams of high-energy subatomic particles such as electrons.

We still haven't reached controlled fusion in this fashion either. Larger, more reliable lasers are needed—or more powerful electron beams.

Still, at the rate we are going now, it seems that sometime before the mid-1980s, one or the other of the methods—magnetic fields, lasers or electron beams—will work. Perhaps all three will work.

Fission Versus Fusion

And how exciting that would be! We have atomic power now in the form of uranium fission, but hydrogen fusion would be much better:

☐ Fission uses uranium and plutonium as fuel—rare metals that are hard to get and handle. Fusion uses hydrogen, easy to obtain and handle.

☐ Fission must work with large quantities of uranium or plutonium, so runaway reactions can take place by accident and cause damage. Fusion works with tiny quantities of hydrogen at any one time, so even runaway fusions would produce only a small pop.

☐ Fission produces radioactive ash, which can be extremely dangerous and may not be disposed of safely. Fusion produces helium, which is completely safe; plus neutrons and tritium, which can be used up as fast as they are produced.

☐ Finally, fission only produces a tenth as much energy as fusion, weight for weight.

Of course, even after we finally attain controlled fusion in the laboratory, it may take as long as thirty years to translate that into large fusion-power stations. There may be many engineering difficulties between a small demonstration that pleases scientists and a large, reliable supply that runs the world.

It may well be 2020, then, before we are a fusion society. It would be wise to conserve oil supplies and to substitute other energy sources (coal, shale, wind, flowing water, tides, hot springs, and so on) to keep us going until fusion can take over.

And we might also strive to develop solar energy, making use of the nuclear fusion power that already exists and that we call the sun. [For more information on alternate sources of energy, see Section VI, below.]

IV. COAL AND NATURAL GAS

EDITOR'S INTRODUCTION

If nuclear power can't solve the nation's energy problems in the short run—and perhaps not in the long run, either—what can? No alternative to oil will provide an easy solution, US energy planners have discovered. To relieve the pressures that the scarcity of cheap oil has brought to bear on the economy, experts are looking for help from a combination of resources that includes coal, natural gas, methane from biomass, oil shale, and solar and wind power. With the exception of natural gas, all these resources are plentiful. Unfortunately, technical, environmental, and economic considerations hamper, to varying degrees, the exploitation of each of them.

Coal is a prime example of an energy source that is abundant yet shackled with constraints, as the first two articles in this section explain. The United States controls at least 435 billion tons of usable coal—enough to meet the nation's needs for more than five hundred years, according to one estimate. US President Jimmy Carter proposed using those resources to reduce the nation's dependence on foreign oil, and in April 1979 he set a production goal of 1.2 billion tons of coal a year by 1985.

Most experts now call Carter's goal unrealistic. Achieving it would require a 60 percent increase in annual production over the 715 million tons mined in 1979. Furthermore, the goal is predicated on an increase in the demand for electricity—and thus the coal to generate it—that shows no sign of materializing. Experts who once figured on a 7 percent annual increase in demand for electric power now expect no more than a 4 percent annual increase during the 1980s. As electricity gets more expensive, consumers use less of it.

Coal, explains John I. Mattill, editor of *Technology Review*, in the first article in this section, seems to be "the ace-in-the-hole that isn't there." Exploiting coal from western states, which hold most of the safer-burning, low-sulphur coal, presents enormous environmental and distribution problems, Mattill notes.

Bob Tippee, district editor of a petroleum industry trade magazine, *The Oil and Gas Journal,* presents, in the following article, the industry's view of the obstacles to increased coal production. Among these obstacles, Tippee cites numerous federal laws that regulate all phases of the industry. Many of these regulations add to the cost of mining and burning coal the cost of protecting the environment and the health of miners and of the general public. Tippee's laundry list of regulations reminds us that the cost of minimizing hazards is hidden in the price of all energy.

Natural gas provides about 25 percent of the nation's energy today. As a widely available fuel, it faces an even more problematic future than coal, but for different reasons. At the present rate of consumption, conventional natural gas—the nation's cleanest, least-polluting fuel—will run out by 1990, less than a century after it was first produced from wells in Texas and along the Gulf Coast. Before the twentieth century Americans used gas made from coal, as they may well do again. (See Section V.)

Natural gas heats about half the homes in the United States, and it is one of the electric power industry's major fuel sources. It powers a wide array of appliances, including air conditioners and clothes driers, where the only substitute power is the far more expensive electricity. Billions of dollars are invested in pipelines that crisscross the nation carrying natural gas from wells and storage tanks to homes and industries. In this section's concluding article, Eugene Luntey—president of a company that owns many miles of pipeline—makes several recommendations. Recoverable gas has recently been discovered in reservoirs under the floor of the Atlantic Ocean, and it is also known to exist in great quantities in the East in rock called Devonian shale and in the West in thick layers of sand known as "tight sands." Luntey recommends tapping these sources and supplementing them with

methane made from manure, plants, and other forms of biomass. (For more on biomass, see H. L. Breckenridge's article
in Section VI, below.)

COAL: THE ACE-IN-THE-HOLE THAT ISN'T THERE[1]

According to the national energy plan, coal is to assume
the role of America's premier fuel in the 1980s.

Heavy reliance on abundant, clean-burning western coal
is likely, and plans have been discussed to transport it by rail
or by pipeline (as a slurry in either water or oil) to points of
consumption in south and east. Alternately the coal could be
burned at the mine or converted there to a synthetic gas and
used to generate electricity for baseload consumption.

Unfortunately each of these alternatives seems fatally
flawed to someone. Energy, transportation, and environmental policies are so much in conflict that the "coal-fired economy" has little resemblance to a "reasoned national decision," says William F. Lipman, a Washington specialist in
public policy studies. He told the American Association for
the Advancement of Science this winter that national policies
"reflect a logistical and environmental myopia."

For example, transporting vast quantities of coal—on the
order of a billion tons per year eventually—from west to east
by rail is the currently favored alternative. The railroads are
confident that the business of moving as much as 60 million
tons of coal a year by 1985 assures them a prosperous future.
But critics point out that to move so much coal would require
over thirty 100-car unit trains a day, and to some communities in Colorado this prospect of two or three loaded and as
many unloaded trains an hour is bad news. The pollution
from autos idling at grade crossings added to that from diesel
locomotives will violate air quality standards; over 140 rail-

[1] Reprint of article by J. I. Mattill, editor. *Technology Review.* 81:68. Mr. '79.
Reprinted by permission. Copyright 1979 by the Alumni Association of the Massachusetts Institute of Technology.

road-highway grade separations—at $1 million each—will be needed by 1985, according to Jack Kinstlinger, Executive Director of the Colorado Department of Highways, whose budget is orders of magnitude inadequate to meet that need.

Most surveys say the railroads themsevles can do the job. But Mr. Lipman and his colleague Aaron Gellman aren't sure. They think the railroads may need $7 billion to $10 billion of new capital to get ready, new technology—and the creative management to use it.

The railroads will also have to figure in managing the ash and sludge left over after coal is burned. A 1000-megawatt plant fired with Wyoming coal would generate about one acre-yard of sludge from its scrubbers daily and 200,000 tons of ash and 600,000 tons of fly ash annually. There is "better than a small chance" that these wastes will contain enough heavy metals and radon gas to be classed as "hazardous" under federal regulations, said Mr. Lipman, and no one seems to know what will be done with it.

If coal can't be moved by rail, what about slurry pipelines? That question seems to lead to many more—and to very few answers. Will there be enough water? What about the quality of water after it's taken the coal from Wyoming to the Midwest or South? How will the coal be dried after transporting? Or burning it wet? What effect will freezing temperatures have on the slurry? Could oil (perhaps synthetic oil) be used as the slurry liquid?

A similar array of difficulties confronts advocates of other coal-utilization strategies. A commercial process for on-site coal gasification may be possible by the "late 1980s," say Antarn P. Sikri and Edward L. Burkwell of the US Department of Energy's Division of Fossil Fuel Extraction; but questions about subsidence and ground water contamination were raised by Professors Harold L. Bergman and G. Michael De-Graeve of the University of Wyoming. More conventional systems for converting coal to electricity at the mine-mouth will probably run afoul of Environmental Protection Administration standards for maintaining the existing high quality of air in mine areas, thinks Mr. Lipman, and they depend on new technologies—mostly untested—for transmitting high-voltage, direct current electricity.

Policy and technology together compound uncertainty and create what Mr. Lipman calls a "pervasive hesitancy" in industry and investors to look favorably at a future based on coal. What's needed, said Mr. Lipman, is "a level of analysis, a commitment to long-term planning and development, and a degree of intergovernmental collaboration that has yet to emerge—anywhere."

COAL'S PROBLEMS[2]

Coal is a potential US energy giant caught in a Lilliputian web of regulations, labor problems, and sputtering demand.

In terms of supply and demand, it's the perfect obverse of petroleum. While the US oil and gas industry must strain to meet soaring demand with production from shrinking reserves, the coal industry sits on a continent full of coal in a country hesitant to mine and burn the stuff. If problems confronting increased coal use can't be overcome, oil and gas will have to strain even more.

Oil and gas make up about 7 percent of total US energy reserves and fill about 75 percent of the nation's energy needs. By contrast coal accounts for about 80 percent of US energy reserves and produces less than 20 percent of the nation's total energy.

Moreover, coal demand hasn't been spurred by rising oil and gas prices as much as might be expected. Especially since the Arab embargo of 1973, coal's role as a petroleum substitute seemingly should have mushroomed. The petroleum industry, with increasing coal production of its own, generally supports such a role.

[2] Article entitled "Coal's Problems to Keep Pressure on Oil and Gas," by Bob Tippee, district editor. *Oil and Gas Journal.* 77:47–51. Mr. 26, '79. Copyright 1979 by the Petroleum Publishing Company. Reprinted by permission.

Yet, since the embargo, coal consumption in the United States has increased only 2 percent to 620 million tons in 1978 from 556 million tons in 1973.

It's not for lack of official goals. Every Administration since the embargo has called for increased US reliance on coal. The Power Plant and Industrial Fuel Use Act passed last year [1978] called for conversion of some existing power plants to coal from oil and gas. And it prescribed use of non-petroleum fuels—presumably coal to a large extent—for new power plants.

Meanwhile, the Government has set ambitious—but feasible, says the coal industry—production targets: 1.2 billion tons/year in 1985 and 2 billion tons/year in 1990, which would be about 28 percent of total US energy output.

But there are obstacles galore, many of them erected within the very Administration that's pitching loudly for increased coal use.

About the only regulatory snag coal producers don't share with their petroleum counterparts is price control. In place of that nightmare is a snarl of often-conflicting safety and environmental regulations that make production difficult and detract from coal's desirability as a fuel.

Furthermore, the fuel use act has been riddled with so many exemptions that some analysts wonder whether it really will result in large-scale conversion to coal from oil and gas.

"We are sitting on billions of tons of coal reserves that we can neither dig nor burn because of a tangled mass of bureaucratic red tape," says R. E. Samples, chairman and chief executive officer of Consolidation Coal Company.

Compounding the regulatory problem are labor troubles and declining productivity.

The bottom line to coal's potential as an energy resource, however, is demand. New mining, land reclamation, and transportation methods can overcome production impediments. But they must do so without increasing price of the product beyond prices of competing fuels.

And there's still the problem of whether potential users will choose to convert to coal—or to synthetic liquids or gas derived from coal—in the face of soaring costs for conversion or compliance with environmental laws.

"The coal industry is not presently limited by its ability to supply," says Samples. "The technology is there, the manpower is there, the reserves certainly are there. What is not there—at least to the extent it could be there—is demand."

The Potential

The coal industry says US coal reserves could satisfy demand for the product for the next five hundred years. Reserves estimates seem to bear that out.

According to the National Coal Association (NCA), known recoverable coal reserves in the United States total 218.4 billion tons. Known recoverable reserves are computed at 50 percent of demonstrated recoverable reserves—known or indicated deposits that are economically recoverable with present technology.

Estimates of ultimately recoverable US coal reserves range from 1.04 trillion to 1.79 trillion tons.

Although large coal producers and trade groups say the industry can meet national production targets of 1.2 billion tons/year in 1985 and 2 billion tons/year in 1990, the US Government isn't quite so optimistic. . . .

[In 1978] the Department of Energy forecast 1985 output of up to 1.188 billion tons/year and 1990 output of up to 1.857 billion tons/year. . . .

On the shorter term, bituminous coal output this year [1979] is expected to hit a record 713 million tons, according to NCA. US production last year totaled about 646 million tons, compared with 688.6 million tons in 1977. Output expected for this year still would be less than estimated productive capacity of more than 800 million tons/year.

Tapping just half of that spare productive capacity could bring about an immediate reduction in oil consumption, NCA says. Earlier this month [March 1979], NCA Chairman Robert H. Quenon, president of Peabody Coal Company, told [former] Energy Secretary James Schlesinger that "a relatively modest increase" in coal use by existing utilities could achieve as much as one half the Administration's goal of a 1 million b/d [barrels per day] decrease in oil consumption.

"Specifically, we estimate that a savings potential of between 250,000 and 500,000 b/d could be achieved on an oil equivalent basis without installing any new coal-fired capacity," Quenon said.

His plan calls for increasing coal utilization rates of coal-fired power plants in the Middle Atlantic and Midwest states and transferring the extra power to the oil-dependent Northeast. It also would require that some Clean Air Act requirements be eased to allow some units now burning oil to convert to coal.

The Obstacles

Of all the problems facing the coal industry, the regulatory maze is by far the biggest. In simplest terms, regulations covering land use, safety and health, and coal consumption drive up costs, making it increasingly difficult for coal to compete with other fuels.

Most additional cost imposed by regulatory requirements is reflected in coal price, both at the mine mouth and at the point of use [says a Library of Congress study on coal, prepared for the Senate committee on governmental affairs].

Coal is in competition with other fuels in all its markets. Increases in the cost of using coal will reduce the incentive for increased coal use and, hence, for reduced oil and gas use.

Last September [1978] Peabody's Quenon made what may be the industry's first attempt to express the effects of regulations and labor settlements in terms of dollars and cents.

In a letter to Robert S. Strauss, Carter's special counselor on inflation, Quenon said:

The combined impact of the 1978 United Mine Workers' wage settlement, black lung excise tax, Surface Mining and Control Act of 1977, and Federal Safety and Health Act of 1977 alone, exclusive of normal materials and supplies cost increases, raises Peabody's production costs in 1978 about $3.75/ton over a 1977 average price of $12.99

He broke out these factors in the cost increase:

☐ *Labor settlement.* Settlement of the United Mine Workers of America strike in March 1978 included a 20 percent salary increase last year and a 38.6 percent increase during the three years covered by the contract. Quenon cited an estimate by the Bituminous Coal Operators' Association (BCOA) that the settlement would mean an average increase of $2.83/hour in wages and fringe benefits in 1978.

"Using the 1977 average output of BCOA-operated companies of 12 tons/man-shift, this translates to a $1.89/ton increase. For Peabody, which has a higher productivity than the industry average, it will mean an average of about $1.19/ton," Quenon said.

☐ *Land reclamation.* Costs of compliance with the Surface Mining Control and Reclamation Act of 1977 will add an estimated $2.50/ton for coal from surface mines (about one-half of US production) and 30¢/ton to the cost of coal from deep mines.

☐ *Safety legislation.* Regulations implementing the 1977 Federal Safety and Health Act weren't in place when Quenon made his estimate. But he said early studies indicated costs of complying with the law would be 18¢/ton for coal from surface mines and 90¢/ton for coal from deep mines. In addition, beginning last April 1, the industry began paying black lung tax of 25¢/ton for coal from surface mines and 50¢/ton for coal from deep mines.

Mining Regulations

Regulations governing mining operations fall into the general categories: land management and miner safety and health.

Among land management regulations are several familiar to the petroleum industry. These include the Wilderness Act of 1964, the Federal Land Policy and Management Act of

1976, the National Forest Management Act of 1976, and the Alaska Native Claims Settlement Act of 1971.

There are additional laws that relate more specifically to coal, including the Federal Coal Leasing Amendments Act of 1976, the Mining in the Parks Act, and the Surface Mining Control and Reclamation Act. Of those, the leasing amendments and surface mining acts pack the biggest punches.

The leasing amendments act requires that "comprehensive" land use plans be prepared prior to leasing of coal resources on federal lands. Where the Federal Government owns the mineral rights but not the surface, lease sales can be held if the lands are part of comprehensive land use plans prepared by the state or part of a "land use analysis" prepared by the Secretary of Interior.

The Surface Mining Control and Reclamation Act prohibits surface mining in several categories of federal land and imposes strict standards for restoring land to original conditions after mining.

Final regulations implementing the surface mining act impose further controls on strip mining. The NCA and several large mining companies failed in their attempts to block the interim rules. Final regulations, proposed last September [1978] were adopted earlier this month [March 1979].

The full effect of the law hasn't been felt. The act established the Office of Surface Mining [OSM] to administer the law in cooperation with the states. Funding problems and legal challenges early in the agency's life have set it behind schedule. According to the Library of Congress study on coal prospects, "This delay has had chaotic effects for the surface mining industry."

There have been challenges from the coal industry and from some members of Congress that the OSM has taken an excessively hard line in attempting to control surface mining.

"Many within the industry and some members of Congress who favored initially the regulation of surface operators now feel that OSM is overzealous in its approach to regulation, which is reflected in the sheer bulk of the detailed requirements," says the Library of Congress study.

Safety and Health

Miner safety and health is a major swing issue in the coal industry's characteristically stormy labor-management relations. In general, managers say safety legislation, specifically the Federal Coal Mine Health and Safety Act of 1969, increases costs and cuts productivity.

Consol's Samples describes the act as

an all-encompassing piece of legislation that was passed in an emotional atmosphere following a tragic coal mine disaster. This law has changed the very nature of mining. And although it has been successful in reducing coal mine fatalities, it has not had a significant, positive impact on nonfatal accidents.

Too many of its provisions have served only to raise costs and reduce productivity while not improving safety.

In a House hearing in 1977, Samples' predecessor, Ralph E. Bailey, said that in a large underground mine, 15-20 percent of the employees are engaged in the performance of tasks related to the coal mine safety act.

Coal industry labor representatives tend to regard productivity considerations as management disregard for employee safety. Still, productivity has declined since the act was made law, falling about 70 percent from the peak year 1968 to 1977.

Has the safety act made coal mines any safer? In general, yes.

Fatality frequency in bituminous coal and lignite mines has decreased to 0.34 fatalities/million man-hours in 1977 from 0.93 fatalities/million man-hours in 1967, according to the Library of Congress. The frequency of disabling injuries declined to 37.5/million man-hours in 1977 from 42.3/million man-hours in 1967.

The accident rate decline has been more pronounced for bituminous underground mines than for surface mines. Safety indicators have been sporadic for anthracite mining, which makes up just a fraction of the modern coal industry.

The 1969 safety act also established the Black Lung Benefits Program, which provides monthly payments and medical benefits to coal miners totally disabled from pneumoconiosis—black lung—and their dependents. The program has been made permanent and expanded since then.

Some coal operators think the black lung program has become too broad. Samples, for example, calls the 1977 legislation expanding the program "a radical liberalization of what was already a liberal program.

"The program has become, in effect, a miner's pension program rather than a true workmen's compensation program."

Coal mining, especially strip mining, also has fallen under strict environmental controls imposed by the Clean Air Act and Clean Air Act amendments of 1977. In terms of how they affect the coal business apart from other industries, however, those laws weigh more heavily on use of coal as a fuel.

Coal Use Regulations

At the core of debates over coal use is the issue of sulfur dioxide emissions. The Clean Air Act, amendments to the act, and the Environmental Protection Agency [EPA] have generally treated such emissions as harmful pollutants—a much criticized characterization.

The debate began in 1971, when the EPA set sulfur dioxide emission standards that industry, in general, charged were too restrictive. During the winter of 1972–73, EPA circulated a study called Community Health and Environmental Surveillance System (Chess), which supported the tight controls on sulfure dioxide emissions.

Validity of the report has been challenged on a variety of fronts, leading to a 1976 House investigation into charges that one of the report's authors distorted findings of the Chess studies. The author was exonerated, but the hearings committee said he may have "overinterpreted the data."

Whether or not sulfur dioxide is harmful to human health and whether it contributes to levels of sulfate in the air has been an emotional issue ever since. Nevertheless, users of high-sulfur coal have a limited number of options in complying with the law.

They can switch to low-sulfur coal, which generally comes from western mines and has a lower heating value than high-sulfur, eastern coal. They can install "best available technology" for lowering sulfur emissions, which means expensive scrubbing equipment. Or they can try to find another fuel.

Consol, a major producer of high-sulfur coal, earlier this year launched a major effort to ease sulfur-dioxide emission standards. It filed a petition with the EPA seeking immediate relaxation of the standards and a general reassessment of sulfur dioxide as a pollutant, with particular attention to whether it poses a serious health hazard. EPA has rejected the petition.

The company's message was summed up last year [1978] in a speech by Executive Vice-President William N. Poundstone. He cited an Electrical Power Research Institute estimate that by 2000, the United States will have spent $200 billion for scrubbers enabling the electric utilities industry to meet source performance standards.

"That $200 billion will come out of the consumer's pocket," Poundstone said. "It is a grossly high price to pay for health benefits that are, at best, uncertain and, at worst, may be totally nonexistent."

Labor

The coal industry has an unrivaled record of work stoppage due to labor disputes.

According to the Bituminous Coal Operators' Association [BCOA], coal industry wildcat strikes caused loss of more than 10 man-days/100 man-days available during 1977, when the latest major United Mine Workers of America [UMWA] strike began. The rate had risen steadily from slightly less than 3 man-days lost/100 man-days available in 1973.

By comparison, the average man-days lost to wildcat strikes per 100 man-days available for all US industries was less than one during 1973–77.

"Such instability has caused union coal to suffer competitively in terms of cost but, more importantly, has signaled major consumers that UMWA mines do not represent a reliable source of coal," BCOA said at the beginning of negotiations during the 1977 strike.

The Library of Congress echoed that assessment. Its study cites Bureau of Labor Statistics figures showing an average 947 coal industry work stoppages/year during 1970–76, compared with 184 stoppages/year during 1960–69 and 314/year during 1950–59. Stoppages averaged 500/year during the earlier period of high strike activity, 1943–52, although the shutdowns affected more workers and more working time was lost.

"The sense of instability of supply engendered by many small strikes in the industry may lead some consumers of coal to shift to other fuels, or to nonunion sources of coal where available," the Library of Congress says.

The study pointed out that frequency of wildcat strikes declined after the 1977–78 UMWA strike was settled and attributed the improvement to adjustments in bargaining and ratification procedures.

Demand Patterns

At present, coal has two primary markets—boiler fuel and steel, with about 90 percent of the boiler fuel market made up of electric utilities. Its increased use as a fuel depends on expansion of existing markets and development of new markets, in each of which coal must compete with other fuels.

Areas of potential growth include electricity generation, industrial boiler fuel, synthetic fuels, and chemical feedstocks.

The area with highest potential growth is in electricity generation, by far coal's biggest current market, which used about 500 million tons of coal . . . [in 1978]. As a utility fuel, coal competes mainly with nuclear power because gas use has been restricted by law and rising prices have made oil less attractive.

Coal now fuels about 47 percent of total electricity generation, according to the Library of Congress. About 45 percent of production capacity to be added through 1986 would be coal-fired, which would push coal demand in this sector to about 800 million tons/year.

But that projection is based on an overall demand growth rate for electricity of 5.7 percent/year, which may be high. The growth rate had risen about 7 percent/year until the early 1970s. Recent low growth rates compared with high economic growth rates may signal a permanent electricity demand slowdown.

But nuclear plants also are running into problems, which could give coal a greater share of the overall electricity generation market. [See "What Went Wrong" by C. L. Wilson and "Nuclear's Cost: Too High?" by Richard Munson, both in Section III, above.]

According to the Library of Congress study, "The largest single factor dictating future increases in coal demand will be the electricity demand growth rate. Coal price, transportation costs, and regulatory problems, should they begin to increase more rapidly than the equivalent factors for nuclear power, would reduce the proportion of coal used in new capacity."

As an industrial boiler fuel, coal has had to compete with oil, which it hadn't been able to do successfully until recently except in facilities of 300 MMBTU/hr heat input or larger [MM-10^{12}]. Postembargo oil prices began to make coal more attractive for smaller boilers until the Clean Air amendments of 1977 made it more costly to burn coal.

More than 90 percent of existing industrial boilers can't burn coal at all. So growth of coal demand as an industrial boiler fuel depends on how rapidly industry adds new boilers and whether it views coal as an economic alternative to oil. Neither of those conditions portends rapid growth for coal demand as an industrial fuel in the absence of forced conversion.

Synthetics from Coal

Recent technological advances notwithstanding, commercial production of synthetic gas or liquids from coal isn't expected before 2000. Economics of these fuels are even more questionable.

Recent increases in spot crude prices have sparked optimism among some researchers.

"When imported crude gets to $30/bbl ... we can produce solvent-refined coal at less than the import price," says Duane R. Skidmore, an Ohio State University researcher. Still, commercial production of the product is twenty years away, he said. [Early in 1980 OPEC prices ranged higher than $30/bbl.—Ed.]

About the only hope for nearer term commercial production of synthetics from coal is for low- or medium-BTU gas used near the mine. [See following article, "What's Wrong With Gas?" by Eugene Luntey.]

"There is little chance of any significant market for coal for synfuels production for a decade and a reasonable chance only for a couple of demonstration plants by the year 2000," says the Library of Congress. "Thus, the coal industry is counting only on a demonstration plant market of perhaps 5-10 million tons/year."

A related market may develop for medium-BTU gas as a chemical feedstock.

Product yields aren't as great as those derived from natural gas, but the coal-based product may become increasingly economical as natural gas prices climb. [See Section V, below.]

The Outlook

Prospects for near term growth in coal demand seem best in the electric utility industry. In the longer term, gasification and liquefaction likely will play increasing roles as prices for gas and oil increase.

Supporting coal use in electricity generation, in spite of environmental rules, are plans for 220 coal-fired generating units to be built between 1978 and 1987. According to Consol, that will increase coal demand by 350 million tons/year.

But measured against the vast US supplies—which are about 30 percent of total world supplies—that seems shy of potential.

J. Allen Overton, Jr., president of the American Mining Congress, summarized the paradox of coal potential at a meeting of the Association of Professional Geological Scientists:

You and I know that America is the Saudi Arabia of coal, and the more we extract it the less we'll have to keep bowing to Mecca for oil.

Perhaps in the long run nuclear fusion or solar power or some other esoteric form of energy will ride to our rescue.

But, between then and now, we need a resource that will bridge the gap. And the name of it is coal.

WHAT'S WRONG WITH GAS?[3]

The United States is in dire need of a new energy direction. "Foundering" is a kind description of our current energy policy. Devaluation of the dollar, inflation, and a dangerous military situation can be traced directly to the high level of oil imports. Other nations know this, and our refusal to act responsibly on energy policy is creating a worldwide lack of confidence. But there is a practical, logical, and environmentally acceptable alternative to this policy—a national commitment to domestic methane.

Suppose President Carter were to say to the world: "This nation is dedicated to emphasizing energy supply from domestic methane. We are going to utilize our vast domestic natural gas resources and develop future methane supplies from coal and all other sources." If this were to be an all-out national effort, I believe that natural gas usage could be increased by 20 percent in two years. This would mean a 20 percent reduction in the import of foreign oil, and a reduction of about $10 billion in payments to foreign nations. (Doubling the amount of methane now used in the United States would eliminate the need for foreign oil entirely.) This

[3] Reprint of article entitled "Delivering Methane," by Eugene Luntey, president, Brooklyn Union Gas Company. *Environment.* 21:33–6. Ja. '79. Copyright © 1979 Helen Dwight Reid Educational Foundation. Reprinted by permission.

scenario is not fiction but a very definite possibility. A substantial body of evidence indicates that a workable, practical answer to the energy problem lies literally at our feet—methane, a permanent source of energy.

The natural gas we currently use is, of course, mostly methane. But natural gas is by no means the only source of this clean-burning, environmentally acceptable fuel. Methane can be made from coal, from peat, from garbage or sewage, and from plant material of many types. It is important to start thinking about sources of methane, since so many people in Washington are saying that "natural gas is running out."

Almost all our energy policy is based on the assumption that there will be a decline in the use of gas as a major energy source. However, there is increasing evidence that gas is not running out. In fact, the supply of methane can be greatly increased. Even fossil methane (natural gas) is not running out. There are still huge quantities of naturally occurring methane to be found, and methane from renewable sources can be available forever. Future generations need not be deprived of the advantages of methane, because the methane made from coal or peat or sewage or biomaterial is exactly the same as that recovered from natural gas wells. All of the advantages of natural gas can be realized for centuries simply by *producing* methane. And it is important to remember that a widespread natural gas pipeline and distribution system already exists and can be used as the methane delivery system. Methane from any source will burn equally as well as natural gas in present and future gas appliances and equipment. A greatly increased volume of methane energy can be delivered by the existing system, which is presently underutilized.

Methane for the Future

The projection for future gas usage in most Government estimates is for a leveling-off or a decline in future years, while electric usage is projected to increase substantially. Coal is also projected to be a rapidly growing energy source. I believe these estimates are incorrect and that decisions based on them will be wrong environmentally, dangerous militarily, and extremely expensive. Such estimates offer no possible promise of quickly affecting the major economic and military danger to the nation—our excessive dependence on imported oil. Worst of all, these predictions will be self-fulfilling because of the huge Government subsidies and research grants which are being directed toward these alternatives.

The indications are that, given adequate economic incentives, large amounts of natural gas could be produced. Consider, first, that there has never really been a concentrated search for methane in this nation. The major gas producers are the major oil companies. By their very nature they are liquid-oriented. Most of their investment is in oil refineries, tankers, oil pipelines, oil or gasoline terminals, and gasoline stations. When they discover gas, it is sold to a pipeline company and that is the last the oil companies see of it. Oil is their business from start to finish. There is nothing wrong with this, but it does explain why methane (natural gas) has not been a primary target for their exploration efforts.

Secondly, the recent discovery of methane in the Atlantic Outer Continental Shelf points up the fact that there are large areas not yet explored for gas. Only a small fraction of the potential production area has been tested. In addition, there is natural gas to be found in deeper horizons in existing fields, in marginal areas near existing pipelines, in tight formations of the Rocky Mountains, in the widespread Devonian Shale deposits of the East, in coal seams, and in deeper water areas not yet tested. All have great potential for methane.

The recent discovery in Canada of possibly the largest methane field in North America clearly points out the inadequacy of the search for fossil methane. There are huge volumes in the earth awaiting either discovery or technology or both to make them available. And the delivery system is already there waiting to deliver this clean energy to consumers nationwide.

When one thinks "methane," one is no longer limited to natural gas. There are vast deposits of coal in this nation but there is no widespread coal delivery system. It is far more ef-

ficient to produce methane from coal and deliver it through
the existing pipeline system than to use it in any other environmentally acceptable way. Coal-to-methane-to-the-con-
sumer makes a great deal of sense. Peat, too, can be used for
gasification and has many advantages such as ease of mining
and high hydrogen content.

Garbage and sewage disposal, which are major urban
problems, can also become part of our system of energy production. The natural product of the decay of organic material
in the absence of oxygen is methane. The potential is there to
solve refuse disposal as well as energy problems. Gas-burning
appliances do not care where the methane originates; they
operate in the same responsive, clean-burning manner on
methane from any source. Methane from waste should be part
of the future supply.

One of the most practical ways to use solar energy is to
produce methane from vegetation, a direct product of the
sun's rays. This vegetation could be grown anywhere but
would grow most rapidly in the southern part of the nation or
in the oceans. Kelp, water hyacinths, or any fast-growing crop
could then be converted to methane, put into existing pipelines, stored in underground exhausted gas wells, and used
when needed. No other method of utilizing the sun to produce energy combines the advantages of regional production,
storage, and widespread availability for use in existing homes
and businesses.

Another possible large energy source may be the geopressured zones known to be present in the South. These salt-
water deposits (8,000 to 22,000 feet deep) exist at very high
pressures and temperatures and are thought to be saturated
with methane. If this water could be brought to the surface,
the methane released, the pressure used for electric generation and the temperatures used to produce fresh water by
evaporation, a new energy source of tremendous potential
would become available. Methane from this source could also
be distributed by the existing gas delivery system. The National Academy of Sciences' estimate was that there might be
more energy in the form of methane in these deposits than

there is energy in all the coal in the nation. The Department of Energy's estimate is much lower but indicates this source could produce one hundred years' supply at existing natural gas usage rates. The potential of this promising source of domestic energy has never been tested.

So methane from conventional natural gas wells, from new areas never explored, from deeper horizons and tight formations, and from geopressured areas, from coal and peat, from garbage, sewage and biomass, are all possible sources of quick, efficient, clean, and economical energy because of the versatile, underutilized, expandable methane delivery system already in place.

Advantages

A big advantage of methane is that existing equipment and appliances do not need to be replaced. Gas will efficiently fire almost any appliance, whether it was designed for oil or gas. Methane and solar systems of energy production make an ideal marriage. Conventional solar equipment usually requires a source of auxiliary heat. Gas is already connected to most places which might use solar, and methane is an ideal automatic back-up fuel, which is available to provide additional energy whenever required. And the regulated gas utility industry could easily offer future solar users the maintenance services needed for the integrated heating and cooling systems of the future.

The methane delivery system currently consists of 350,-000 miles of underground high-pressure pipelines, 650,000 miles of distribution mains, and connections to 45 million energy users (nearly three fourths of the people in the nation use natural gas in some way). This delivery system now carries much less gas than it could carry because of the regulation-caused shortages of recent years. The delivery system is also connected to an efficient storage system of exhausted underground gas wells into which methane from any source could be pumped and then recovered as needed. This storage system could be readily expanded and could thus be used for strategic storage of methane. Enough methane could be stored to replace foreign oil if the foreign supply should be interrupted; and methane can be quickly delivered nationwide, whereas strategic oil storage requires unusual and expensive provisions for delivery.

Obstacles

If this proposal makes sense, then why not more emphasis on methane? Why have the projections of the Department of Energy completely phased out gas as an energy source by the year 2030? Why are 80 percent of DOE's research funds devoted to the generation of electricity and only 8 percent to methane, including coal gasification? Why is methane not at least given a chance if it has such potential for the future? Why are all federal housing regulatory decisions made on an extremely narrow basis which ignores the value of existing pipelines as a fundamental energy delivery system?

It is my reluctant conclusion that methane, and its promise for the future, may be regarded as a threat by many established interests. Increased use of methane could make the current large expansion in electric generation unnecessary, and this is an obvious threat to the electric-industrial complex built up around the nuclear development of the past thirty years. The major oil companies are also unlikely to be interested in methane, because they process both domestic and foreign oil. A reduction in the importation of foreign oil could reduce the use of their refinery facilities—even if they produced more natural gas. Another point to consider is that if there is a readily available, environmentally acceptable, economical solution to the energy problem, we might not need a 20,000-person, $12-billion Department of Energy. In addition, Congress, unfortunately, has considered natural gas on the narrow basis of wellhead price and not on the broad basis of an environmentally acceptable, deliverable, continuously available domestic energy supply. The natural gas industry is, and will continue to be, a regulated public utility type of service and is thus not very effective in telling its story against such odds.

These are undoubtedly some of the reasons why present government policy does not even consider the future use of methane as a major energy source. Nonetheless, methane and the methane delivery system must eventually become the fundamental building block of the future energy policy of this nation. There is no practical alternative.

V.   ALTERNATIVE FUELS

EDITOR'S INTRODUCTION

Liquefaction and *gasification*—these two words describe the processes by which coal is changed into other forms. The resulting "synthetic fuels"—though technically not synthetic at all, since they have the same carbon base as crude oil pumped from the ground—are the cornerstone of President Carter's 1979 plan to make the United States less dependent on imported oil. (For details of that plan, see "Phase 2 of Carter's Energy Plan," in Section II, above.) Two other "syn-fuels," shale oil (extracted from oil shale, a rock) and oil trapped in gooey tar-sand deposits, are already being produced in the United States and Canada. So is still another alternative fuel, methanol, distilled from plants and garbage and used in the concoction sold at service stations under the name of gasohol. (Various methods of converting biomass into alcohols are discussed in Section VI.)

Will these alternative fuels help the nation face the challenge of an oil-short future? And, if so, how soon? The two pieces in this section explore answers to those questions. The first piece examines the feasibility of alternative fuels—especially fuels that would replace the increasingly costly petroleum products used by automobiles. The selection is an excerpt from a book produced by staff members of the Worldwatch Institute, a Washington-based research institute devoted to the analysis of emerging global issues. Its authors take a less optimistic view of the economics of synthetic fuel production than do William M. Brown and Herman Kahn, the authors of this section's second piece. Brown and Kahn, members of another research organization, the Hudson Institute, take the view that a firm government commitment to a synthetic fuels program could produce alternative fuels that, barrel for barrel, would cost less than imported oil.

Backers of such a commitment remind us of the crash program that produced synthetic rubber during World War II. In short order, 87 percent of the nation's rubber was factory-made. The right government policies today, some energy experts believe, could result in alternative fuels meeting up to half the nation's energy needs by the year 2000. Whether the nation follows this "hard path" toward the future will most likely depend as much on economics and on how well the technical difficulties are surmounted as on how willing Americans are to accept the possibility of further environmental degradation.

ALTERNATIVE FUELS FOR CARS[1]

On the eve of the Industrial Revolution, England was running out of energy. Wood and charcoal accounted for most of the nation's energy consumption, but many of its once lush forests had been cut down to fuel ironworks, potteries, and other nascent industries. Industrial expansion would have been impossible without the development of alternative sources of fuel. The rapid growth in coal production overcame the wood shortage, however, and the Industrial Revolution took off.

Is it conceivable that, in the same way, petroleum substitutes could be developed swiftly enough to shield the automobile from the impending downturn in world oil production? The short answer is that it is conceivable, but it is unlikely. The leading candidates—alcohol, liquid fuels from coal, and oil from tar sands or oil shale—could eventually be produced in large quantities. But they all face serious economic, environmental, or social problems that are likely to present major barriers to their development.

[1] Excerpted from the book *Running on Empty*, by Lester R. Brown, head of Worldwatch Institute, a "think tank," and by Worldwatch researcher Christopher Flavin and senior researcher Colin Norman. *Running on Empty: The Future of the Automobile in an Oil-Short World.* Norton. '79. p 34–45. Selection is reprinted by permission of W. W. Norton & Company, Inc. Copyright © 1979 by Worldwatch Institute.

An alternative approach—the development of vehicles that do not require liquid fuels at all—is also unlikely to provide salvation from oil shortages. Electric cars are the chief contenders for this role, but difficult technological obstacles must be overcome before electricity can seriously challenge petroleum as an automotive fuel. And more exotic alternatives, such as hydrogen-powered automobiles, are even further from becoming everyday vehicles.

The only nonpetroleum fuel now being used in any significant quantity in cars is ethyl alcohol, or ethanol. In many respects, ethanol is an ideal fuel. It is renewable, for it can be produced from plant materials, such as sugarcane, corn, cassava, and sweet sorghum, from agricultural wastes, and from municipal garbage. It burns cleanly, producing virtually none of the pollutants associated with gasoline or diesel oil. And the technology for making it is relatively simple—it differs little from the centuries-old practice of distilling liquor from fermented grain.

Ethanol in fact has a long history as an automotive fuel. It was used to run some of the earliest automobile engines, and Henry Ford equipped some of his first models with carburetors that could be adjusted to take alcohol, gasoline, or any mixture of the two. Modern engines are less tolerant than their ancestors, but they can run on mixtures that contain up to 20 percent ethanol. Such blends, popularly known as gasohol, were marketed briefly in the United States during the twenties and thirties, and they are now being sold there again as well as in Brazil.

Automobile engines must be modified to run on pure ethanol, but the conversion costs are relatively modest. The changes include carburetor adjustments, the addition of a device to improve starting, and alterations to the cylinder block to raise the compression. According to studies in Brazil, an engine designed to burn ethanol should cost no more to produce than one meant to burn gasoline, but it would cost between $300 and $400 to modify an existing engine.

Leadership in alcohol production is coming, in fact, from Brazil, a country that is heavily dependent on imported oil.

Stung by the 1973 oil price increases, the government launched an ambitious program in 1975 to convert part of Brazil's large sugarcane crop into ethanol. The original goal was to meet 20 percent of the country's automotive needs with alcohol by 1980, and to become self-sufficient by the end of the century. By mid-1979, Brazil was producing almost 700 million gallons of ethanol, about 14 percent of its gasoline requirements.

Flushed with the early success of the program, and hit again by a new round of oil price increases, the Brazilian government announced in mid-1979 that an additional $5 billion will be invested in new distilleries by 1985. In addition, 1.2 million new cars coming off the assembly lines between 1979 and 1985 will be designed to run on pure alcohol, and some 475,000 others will be converted to do the same. Although there is now no official target date, complete self-sufficiency as soon as possible is the ultimate goal.

The only other country producing significant quantities of alcohol to run cars is the United States. Although the leadership in Brazil is coming from the government, in the United States it is largely the result of private initiative. The liquor industry, for example, is reactivating some old distilleries, and about 100 million gallons of alcohol for automotive fuel will be produced in 1979—or roughly 0.1 percent of total US gasoline consumption. Corn is the chief source of ethyl alcohol in the United States at present.

Although the development of ethanol as an automotive fuel is progressing rapidly, it is currently far more expensive than gasoline and the economics of large-scale production are riddled with uncertainties. At early 1979 corn prices, ethanol in the United States could be produced for about $1.50 per gallon, or about $63 per barrel. At those prices, mixing ethanol with gasoline to produce gasohol is, as one commentator put it, "like watering beer with champagne."

The cost difference between ethanol and gasoline could shrink as oil prices rise and as the economies of scale bring the cost of alcohol production down. But there is also a good chance that other forces will push costs in the opposite direction. Corn prices in early 1979 were depressed after four years of bumper harvests, for example; if prices rise in the future, the cost of ethanol will also increase. A study by the Department of Agriculture notes that "natural fluctuations in corn production levels can prove devastating to the cost and availability of ethanol feedstocks for any significant-scale gasohol program." The very success of a large-scale alcohol fuel program could, in fact, boost the demand for corn to such an extent that prices would rise and ethanol would become more expensive. A similar trend could occur in Brazil, as the ethanol program was initiated at a time when the sugar market was depressed and cane prices were low.

Both Brazil and the United States are using tax policies to encourage the use of alcohol and gasohol. In June 1979, for example, although alcohol cost about twice as much to produce as gasoline, it was selling in Brazil for $1.04 per gallon compared with $1.52 for gasoline. In the United States, gasohol has been exempted from the federal gasoline tax of 4¢ per gallon; some states, such as Iowa and Nebraska, have also exempted it from state gasoline taxes of about 6¢ per gallon. These exemptions constitute major subsidies for alcohol production. Since gasohol in the United States contains 10 percent alcohol, a 10¢ total tax exemption per gallon of *gasohol* amounts to $1.00 per gallon of *alcohol*—which works out to be $42 per barrel. It is an expensive way to supplement $20-per-barrel imported oil [as of January 29, 1980, "official" OPEC prices ranged from $24.51 to $34.72 per barrel, with spot market prices $10 higher.—Ed.]. These subsidies, not surprisingly, have provided a great boost to gasohol availability in the United States. By mid-1979, gasohol was being marketed through some eight hundred filling stations in twenty-eight states; in Iowa, it accounted for 2.5 percent of total gasoline sales in March 1979.

The ultimate potential for ethanol as an automotive fuel is likely to rest on the availability of raw materials as well as on price. Brazil is fortunate in having a large sugar surplus that can readily be converted into alcohol. Few other countries have that luxury. Large-scale use of crops around the world to produce alcohol could thus raise serious concerns about competition between food for people and fuel for automobiles. Although Brazil has extensive areas of land that could be used to grow energy crops, for example, it also has many impoverished people who need additional land to grow food on. And although the United States does have a substantial grain surplus that could be converted into alcohol, diversion of a significant part of the surplus could result in less food being available for export in lean years. The purchasing power of American motorists could, in effect, override the basic food needs of people in the developing world.

A Department of Energy review of the potential for alcohol fuels in the United States, published in mid-1979, suggests that if all the land idled under farm programs were brought back into production, and if large quantities of organic wastes such as cheese whey, citrus pulp, and municipal garbage were used to make ethanol, there would be sufficient raw materials to produce about 4.8 billion gallons of ethanol per year. That works out to less than 5 percent of current US gasoline consumption. The review suggests, moreover, that production is unlikely to reach more than 600 million gallons a year by 1985, which would fill only a tiny fraction of automotive fuel needs.

In areas where raw materials such as agricultural wastes and food processing residues are cheap and readily available, ethanol could play an important role in meeting local fuel needs. Efficient small-scale distilleries that use waste wood, corn stalks, or sugarcane residues for heat are likely to provide the key to economic local production of ethanol. So far, however, the Brazilian and American programs have concentrated mostly on large-scale facilities.

In the long term, the development of technologies to convert cellulose materials such as wood, paper pulp, and municipal solid waste to ethanol could greatly extend the resource base. Use of such materials, moreover, would not encroach directly upon food supplies. But methods to convert cellulose materials into ethanol are currently more expensive than those using sugarcane, corn, and other crops, and the Department of Energy suggests that "development of cellulose alcohol processes has not proceeded far enough to justify immediate investment in full-scale plants." Wood and woody products may, however, become important for the production of a different type of fuel: methyl alcohol, or methanol.

Methanol, like ethanol, was used as an automotive fuel in the early twentieth century, and it is still used to run some types of vehicles: all thirty-three cars in the 1978 Indianapolis "500" race ran on pure methanol, for it is especially suitable for use in high-compression engines. Unlike ethanol, however, it cannot serve immediately as a gasoline substitute because it corrodes some of the metals used in standard gasoline engines, and it causes a variety of other problems in performance. Changes in engine materials and design would thus be required for methanol to be widely used as an automotive fuel, and the changes would have to be phased in over several years. An alternative approach, being developed by the Mobil Oil Corporation, is to process methanol to produce a fuel that more closely resembles gasoline; but more than two gallons of methanol are required to produce each gallon of gasoline, and the technology has yet to be tested on a large scale. The Department of Energy has concluded that "it is unlikely that methanol fuel will be used extensively before 1990."

Methanol was originally produced by heating wood in a closed vessel and condensing the gases given off—hence its common name of wood alcohol. Large quantities are now produced from natural gas for use as a raw material in the petrochemical industry. In the future, methanol may be produced from a range of carbon-containing substances, such as wood, wood products, and municipal solid wastes, in massive conversion plants costing billions of dollars. The technology is highly complex and similar in some respects to petroleum refining. Although such technologies are now being developed, the most likely source of methanol in the near future will be coal, not wood.

The possibility of producing so-called synthetic fuels from coal, oil shale, and tar sands has recently received much attention, particularly in the United States. In mid-1979, Presi-

dent Carter came up with a crash program to produce 500,-
000 barrels of synthetic fuels a day by 1985, and 2.5 million
barrels a day by 1990. The centerpiece of his proposal is an
$88 billion fund to provide direct investments, loan guaran-
tees, and similar incentives for the construction of synthetic
fuels plants. These funds would come from the windfall profits tax that will be levied on the oil companies if and when the
price of domestic oil is decontrolled. [President Carter began
phasing out controls in May 1979. They will be entirely eliminated by October 1, 1981, when they would have expired by
law anyway.—Ed.]

President Carter called his proposal "the most massive
peacetime commitment of funds and resources in our nation's
history to develop America's own alternative sources of fuel."
Coming at a time when gasoline prices were rising sharply
and lines were building up at filling stations, the plan met
with initial enthusiasm from an exasperated public. As Senator Dale Bumpers [Democrat, Arkansas] put it, "the American people are in a mood to do something, even if it is
wrong." But doubts about the wisdom of proceeding at a
breakneck pace, before some of the key technologies are
firmly established, quickly began to surface. In mid-Septem-
ber, President Carter announced that he would accept a more
orderly program, with the construction of more pilot plants
before moving to large, commercial-scale operations.

Processes for producing liquid fuels from coal were developed a half century ago, but there is still much uncertainty
about which technology constitutes the best and most economical approach. The different processes yield different
products, ranging from liquids that closely resemble crude oil
to more volatile substances such as methanol. They have one
thing in common, however: they all require massive, expensive facilities.

Germany pioneered the development of liquid fuels from
coal during the twenties and thirties, and then put the technologies to use during World War II. By the end of the war,
the country was producing nearly 100,000 barrels a day, using
the fuels to power its air force and its panzer divisions. Once

the war was over, however, German industry went back to refining imported oil, a much cheaper alternative. Today, only one country—South Africa—is operating a commercial coal liquefaction plant.

Highly vulnerable to oil embargoes, South Africa opened a plant for converting coal into liquid fuel in 1959. The facility produces an estimated 20,000 barrels of synthetic oil per day, close to 10 percent of the nation's petroleum needs. A much larger plant was commissioned in 1974 and, following the Iranian government's embargo of oil to South Africa, a major expansion of the plant was announced in March 1979. Scheduled to open in 1982, the $6.7 billion facility will produce enough synthetic fuel to meet about half South Africa's oil needs.

While no other country has a commercial plant in operation, several pilot plants have been built in the United States to test various technologies. A consortium of petroleum firms headed by the Ashland Oil Company is now building a plant to convert 600 tons of coal to 2,000 barrels of industrial fuel oil per day, and Exxon has a plant under construction that will use a different technology to process 250 tons of coal per day, producing synthetic oil and gasoline substitutes. A much larger facility, to test commercial-scale production, will soon be constructed in West Virginia with funds from the United States and West German governments, a consortium of Japanese firms, and the Gulf Oil Corporation, which will design and operate the plant. Designed to convert 6,000 tons of coal per day into 20,000 barrels of oil, it is scheduled for completion in 1985.

One of the attractions of coal as a source of synthetic oil is its relative abundance. The same is true of tar sands and oil shale. Deposits of these fuel sources in Canada, Colombia, the United States, and Venezuela contain far more oil than is left in the oil fields of the Middle East. The great difficulty is devising an economical and environmentally acceptable process for extracting the oil.

Tar sands are a mixture of sand, water, and a thick hydrocarbon called bitumen. They look like black, sticky dirt. Production of synthetic oil from tar sands involves mining the sands with huge draglines, treating them with hot water and steam to extract bitumen, and chemically treating the bitumen to produce a synthetic oil that can be processed in a conventional refinery. Oil from tar sands resembles natural petroleum in physical appearance, but not in price: a commercial tar sands plant in Canada has been producing synthetic oil at between $30 and $45 a barrel.

Commercial production of oil from tar sands is currently limited to Canada, where extensive deposits lie in the Athabasca River basin. It is an area where temperatures drop to minus 50°F. in the winter, turning the ground into a substance resembling concrete, and where the spring thaw creates bogs that can swallow large pieces of machinery. Such conditions have not aided tar sands development. A $1.9 billion facility, built with funds from oil companies and the Canadian government, has managed to produce about 50,000 barrels of oil per day, although it was designed to produce twice that amount; a smaller venture, run by a subsidiary of Sun Oil Company, has been operating at a loss ever since it opened a decade ago. The participants in these ventures are not discouraged, however—with continued backing from the Canadian government, they are planning to step up their operations. And two groups of oil companies, one headed by Shell Canada and the other by Exxon, are also planning to invest up to $5 billion each in large new plants.

The extraction of oil from shale deposits is even more difficult and costly than the production of oil from tar sands. The world's richest oil shale deposits lie in an area around the junction of the states of Colorado, Utah, and Wyoming in the United States, where oil is tightly locked into rock formations just beneath the surface. For years, some analysts have been saying that the time is ripe for production of oil from shale, but so far the technology has remained just beyond the economic horizon.

Extracting oil from shale is an arduous business. It involves mining shale rock, crushing it, heating it in a furnace, condensing the oil, and partially refining it. The waste material—powdery rock—must then be disposed of. An alternative process, developed by Occidental Petroleum, involves heating crushed rock underground and requires the mining and disposal of less material, but it has yet to be tested on a large scale. Oil shale development attracted a flurry of interest in several companies following the 1973 oil price rise, but that interest waned rapidly in the face of enormous technical difficulties and huge capital costs. After building small pilot plants, many firms have withdrawn from development of this fuel source, and no commercial ventures have been attempted. A few larger facilities are on the drawing board, but the companies concerned are mostly awaiting the outcome of President Carter's proposals for financing a synthetic fuels industry before pushing ahead.

The 1979 OPEC price rise has rekindled interest in oil shale development, but many skeptics remain unconvinced that the age of shale oil has finally dawned. Edward Merrow, a Rand Corporation ["think-tank"] analyst who recently completed a study for the Department of Energy, has flatly stated that "a viable oil shale industry is unlikely this century." A central problem is cost.

The costs of producing synthetic oil from coal and shale are highly uncertain. The estimated production cost has remained tantalizingly above the rising price of crude oil during the past few years. In 1972, for example, the National Petroleum Council estimated that oil could be produced from coal for between $7.75 and $8.25 a barrel, while oil from shale would cost $8.29 a barrel. If these estimates were realistic and had held steady, investments in synthetic fuels plants would have been extremely attractive as the world oil price rose. Instead, now that oil is about $20 per barrel, the expected costs of producing liquid fuel from coal has climbed to between $25 and $45 per barrel, while that of oil from shale is between $25 and $35. It is not surprising that industrial investment in synthetic fuels facilities has been modest. [However, with the "official" price of OPEC oil topping $34 a barrel, and apparently going higher, these processes may well become economically viable even within a short time.—Ed.]

While the overall cost of synthetic fuels production may be uncertain, it is clear that the initial capital costs will be enormous. President Carter's proposal to sink $88 billion into synthetic fuels plants is only part of the estimated expense. Industrial investments will be at least as large. A study by the Rand Corporation, moreover, has pointed out that the costs of such huge chemical complexes often greatly exceed early estimates. Cost overruns of 300 percent are typical. Such massive capital outlays will divert investments from other areas, including energy conservation programs that could save far more energy than the synthetic fuels program is likely to produce [see Section VII, below]. Milton Russell, an energy analyst for Resources for the Future, has suggested that "the prospects for anything like 2.5 million barrels per day by 1990 seem to me slim unless as a nation we decided nothing much else is important." [Resources for the Future is a conservation organization that works in education and research on conservation, use and development of natural resources, energy and the improvement of environmental quality.—Ed.]

Although the capital constraints are daunting enough, there are other considerations that could slow the development of synthetic fuels production. Every barrel of shale oil requires the mining and processing of about 1.5 tons of rock, which means that a one-million-barrel-a-day shale oil industry would dig up almost as much material as the entire US coal industry now mines. The demand for water is also a serious problem. Between two and five barrels of water are required to produce each barrel of shale oil, and up to thirteen barrels of water are needed to produce each barrel of liquid fuel from coal. The synthetic fuels industry would be concentrated in the western United States, where water is already scarce and where demands from both agriculture and urban development are growing fast.

Other environmental problems could also present serious obstacles. "These are big, dirty plants. Everybody wants these plants—but wants them someplace else," Robert Hanfling of the US Department of Energy has noted. They are expected to discharge soot, sulphur dioxide, nitrous oxides, and other pollutants, and there is some concern that workers in the plants could be exposed to cancer-causing chemicals. Moreover, a massive shift to synthetic fuels would release large amounts of carbon dioxide, which in turn could lead to a warming of the earth's atmosphere. A panel of scientists reported to the US Council on Environmental Quality in July 1979 that more carbon dioxide is released from the production and combustion of synthetic fuels than from the direct burning of coal, and warned that "if we wait to prove that the climate is warming before we take steps to alleviate the carbon dioxide buildup, the effects will be well under way and still more difficult to control.

[For another view on synfuels, see the following article from *Fortune.*—Ed.]

SYNTHETIC FUELS CAN BE ECONOMIC NOW[2]

The newspapers these days are full of proposals for producing synthetic fuels. President Carter has suggested that we create an Energy Security Corporation [see "Phase 2 of Carter's Energy Plan," in Section II, above] to develop such fuels, and dozens of other schemes are being dropped into the congressional hoppers. The proposals are unanimous in assuming that synfuels cannot now be brought in at prices that are competitive with oil. All of them therefore also assume that enormous government subsidies—some proposals call for as much as $100 billion over the next decade—are necessary. Despite these staggering costs, the proposals are said to be justified by the need to liberate ourselves from the prices being imposed by the Organization of Petroleum Exporting Countries.

We agree about the need but disagree about the cost assumptions. It is our belief that synfuels can be brought in at

[2] Reprint of article by William M. Brown, director of energy studies at the Hudson Institute. a "think tank," and by Herman Kahn, the institute's research director. *Fortune.* 100:110–11. Ag. 13, '79. Copyright © 1979 by William M. Brown and Herman Kahn. Reprinted by permission.

prices that are competitive with OPEC's—that are, indeed, significantly lower than those set by the oil producers at their June [1979] meeting in Geneva.

In this situation, what we believe matters a great deal. If we continue to believe the myth that synfuels are still uneconomic, and will become economic only after further OPEC price increases, then we are ensuring that those price increases will take place. The myth will become a self-fulfilling prophecy.

Some basic propositions about synfuels are not in dispute. The United States has abundant supplies of fossil fuels—especially coal and oil shale—and there is no doubt that these can be converted into oil and gas. Detailed engineering proposals for building huge synfuel complexes have received a fair share of attention in recent years.

The problem is that none of the proposals seem to be practical. It is repeatedly said that, for many years into the future, the cost of producing synfuels will make them unable to compete with petroleum and natural gas. Conventional analyses show that, in 1978 dollars, the new synthetic fuels would cost somewhere between $20 and $30 per BOE (barrel of oil equivalent), with estimates tending to cluster around the midpoint of that range. Even after Geneva, the world price of oil is still somewhat less than $25. [As of January 1980, OPEC prices ranged from $24.51 to $34.72—higher on the "Spot" market.—Ed.]

A Need for Thinking

We believe, however, that synfuels can be marketed at around $15 per BOE—a price at which OPEC would have some large problems on its hands. To bring off this trick, help from the US government will indeed be needed. More important, however, we need a new way of *thinking* about synfuel financing.

We find it easy to envisage a $1-billion plant producing 16 million or 17 million barrels of synthetic fuel annually at a $15 price—i.e., generating revenues of $250 million. Operation and maintenance costs might run to $100 million annually. That would leave $150 million available for capital charges.

In a conventional analysis of synfuel economics, the case for going ahead breaks down at this point—because $150 million isn't enough. That is, it would not cover the debt service, taxes, depreciation, and profits. Ten or fifteen years ago, $150 million would have been more than adequate to cover these charges, and entrepreneurs would have rushed in to finance synfuel (or any other) projects to which the numbers mentioned here were applicable. Today, however, the annual capital charges would have to be around $300 million, which is why the product couldn't sell for less than $25 a barrel.

What has changed in the interim? Part of the answer, plainly, lies in recent government policies that make *any* investment less attractive than it used to be. An investor eyeing synfuels would have to consider the possibility that the fuels, once produced, might be subject to price controls. He would have to worry about the fact that inflation results in higher effective tax rates, i.e., because so much of the income being taxed is "unreal." He would also have to worry about the uncertainties associated with fluctuating inflation rates. He would have to factor in the added costs of environmental and safety regulations.

To offset these and still other risks, the equity investor would demand a higher level of profitability in the project: Where once he would have cheerfully accepted an 8 or 10 percent cash-flow return, he is today uninterested unless he can see the prospect of 15 or 20 percent. And because of inflation, lenders also need higher returns. In the 1950s, investors in long-term high-quality bonds accepted yields of 2 or 3 percent. Today, knowing that they will be repaid in less valuable dollars, they demand 9 or 10 percent.

Let us assume, however, that the government decided to encourage rather than discourage the construction of synfuel plants—not through those twelve-digit subsidy programs that are being mentioned in the press, but through some more modest proposals. These would include support of some demonstration plants; they might even include subsidies for some of the early commercial plants. What is mainly needed from the government, however, is a basic decision that in this industry it will remove the particular discouragements associated with overregulation and, more important, with inflation.

The Return Would Be Real

Assuming that some such decisions were forthcoming, we believe it would be possible to structure a financial plan enabling our hypothetical $1-billion enterprise to produce at that $15 price. The plan would involve capital charges of $150 million or less. The per-barrel charges would, of course, vary with the degree of leverage deemed acceptable and also with our assumptions about investors' required returns. Assuming 75 percent debt financing (which is common enough in Japan) and a cash-flow return of 10 percent on equity, the annual capital charges would be around $5.50 a barrel and total capital charges around $85 million. With a 50 percent debt ratio (which is closer to the American norm) and a 12 percent return, the charges would rise to about $130 million, or $8 a barrel.

Meanwhile, the annual operations and maintenance [O and M] charges for a 16-million-barrel plant could be held in a range somewhere between $5 and $8 a barrel—let us say $7. These figures assume that the Federal Government would make available good sites on federal land, where a lot of the feedstocks (oil shale or coal) are located; the figure also assumes that royalties and other economic rents would be modest. In any case, $7 a barrel for O and M, combined with $8 for capital charges, would make possible fuels at $15 per BOE—the price that gives OPEC a problem.

But why would investors—who have been insisting on returns approaching 20 percent on large capital projects in new industries—be willing here to settle for 10 or 12 percent? Because the returns would be real, not nominal.

Specifically, we would offer lenders an interest rate equal

to 3 percent of the bond's face value. While that figure is substantially less than prevailing market rates, the terms of the offering would guarantee a real return, for both the interest payments and the final redemption price would be indexed to the inflation rate. There would be no taxes on "unreal" income. Under these conditions, investors would no longer feel that they needed 9 or 10 percent interest.

In reality, neither debt nor equity investors need those higher rates now. They demand them because inflation, coupled with traditional accounting and financing methods, has confused them about the cost of producing synfuels. Specifically, they are confusing the initial cost with the average cost over, say, thirty years. In an age of inflation, the two are significantly different. The conventionally calculated $25 price contains a mixture of O and M costs, which rise with inflation, and capital charges, which are fixed—at least, they are fixed when viewed in current dollars. But in constant dollars, the situation is quite different: as the illustration makes clear, the O and M figure is fixed (i.e., it stays abreast of inflation), while the capital charges decline. At the end of thirty years, a 7 percent inflation rate would bring $18 a barrel of capital charges down close to $2 and total charges down to $9.36. Over the entire thirty-year period, the average charge (in 1978 dollars) would still be around $15 per barrel. It is the *initial* charge that is set at $25—and it is this that makes the whole enterprise seem uneconomic.

A Payoff in the Barrel

A variation of the financial plan described here might be appropriate in an international deal. Suppose that the Japanese—to pick one likely prospect—decided to finance a $2-billion US synfuel plant. It might be logical for them to accept their return, not in indexed interest payments, but in barrels of fuel, with the number of barrels (it could be around 35,000 a day) held constant over the life of the plant.

We have, of course, been discussing the financing of synfuels using simple hypothetical examples. In reality, we will need many different kinds of plants to produce many different kinds of fuels: low, medium, and high BTU gases, heavy and light boiler fuels, distillate, naphtha, gasoline, and alcohol—as well as a number of petrochemicals and other valuable byproducts. Most of the proposed plants would produce several of these fuels. In the largest projects, the feedstocks are most likely to be coal or oil shale; but peat, wood, and trash could be used in some situations.

Despite these complications, we believe that a reasonably mature synfuel industry can be established in ten or twelve years. By that time, we should have perhaps ten or twenty commercial-scale plants in existence. The energy industry now assumes that federal subsidies will be available for demonstration plants, perhaps even for some of the commercial plants. It may well be that a certain amount of Federal Government "pump priming" will be necessary to get the industry onto a commercial basis. But there is good reason to expect that, not long afterward, the energy industry could develop the required amount of synfuel capacity without federal help.

The Smallest Bid Wins

Although the government will have to help the synfuel industry get started, it will presumably be interested in minimizing its subsidies. This might be done, for example, through a kind of "auction" system in which the government would set a ten-year schedule for the desired production capacity and then encourage private entrepreneurs to submit competitive bids for the needed federal support; each year, the successful bidder would be the company needing the *least* support.

But these kinds of questions are still remote. The urgent question involves the rate at which we begin to build synfuel plants. And the decisive consideration here may be our ability to grasp the simple fact that such plants already appear to be economic—and competitive with OPEC oil.

EDITOR'S INTRODUCTION

One of the great unknowns in America's energy future is the extent to which solar energy, in its various forms, will contribute to the nation's energy supply. The "solar lobby"— a loose confederation of people committed to convincing policy makers of solar's potential—has suggested that solar energy could supply 40 percent of the nation's energy by the year 2000. Yet, others view solar energy as an impractical solution proposed by dreamers and amateur technologists.

Solar's true potential probably lies somewhere between such high hopes and such low expectations. Solar heating systems alone, some experts suggest, could by the year 2000 replace three million barrels of oil a day—nearly half the nation's projected oil imports. Already more than five thousand buildings in the United States rely to some degree on active solar energy systems for heating and cooling and much energy can be saved by the use of architectural design elements such as better orientation of buildings and proper placement of evergreens and deciduous trees.

The use of solar power to heat space and water mushroomed during the 1970s. Experts say that sales and installations of solar heating equipment during the 1980s and 1990s may amount to more than $100 billion.

But the sun's energy can do more than heat and cool rooms and water, its backers point out. In fact, the term *solar energy* covers a multitude of renewable energy producers—from simple wood-burning stoves to windmills to water power to biomass fuels to solar cells that could convert the sun's rays into electricity and beam it down to Earth as microwaves. This last-mentioned technology is called photovoltaics.

Anthony J. Parisi, the New York *Times* energy specialist, opens this section by describing some of solar energy's "small successes" during the 1970s. At the close of the decade, Parisi found solar advocates "retrenching" but promoting the sun as "a serious alternative source of energy" ready to be tapped now.

Most of us have enjoyed solar energy released from biomass on a wintry day, as we stood before a wood-burning fire. But wood is only one form of biomass, as H. Leon Breckenridge, manager of EG&G Idaho Inc.'s Economic Analysis and Biomass Energy Systems, reminds us in the next article, from *USA Today*. "If it grows," promoters of biomass as an energy source say, "burn it—or convert it to energy." Since plants store energy from the sun, many people regard them as ideal solar collectors.

The power of the wind (air set in motion by the sun's heat) is harder to capture than other forms of solar energy, and for this reason wind-generated electricity can be expensive. Still, as freelance writer Lee Stephenson explains in this section's third article, government agencies and private companies are working hard to find ways that will make wind power competitive, economically, with conventional sources of power.

Another type of solar electric energy is hydropower: water evaporated by the sun and deposited on high ground as rain flows in rivers toward sea level, powering turbines as it moves. Hydropower generated a third of the nation's electricity sixty years ago. Today, water power contributes only about 13 percent of the nation's electricity and meets 4 percent of the nation's total energy needs—about the same proportion as nuclear energy. Yet, as electrical engineer George Erskine tells us, in the next article, which first appeared in *Environment*, hydropower will make a comeback if enough small dams can be restored to working order. By one estimate, all the nation's small dams could be repaired and outfitted with the latest electricity-generating equipment for the cost of two and a half nuclear plants. Altogether, those restored dams would produce as much electric energy as sixteen nuclear power plants.

If the nation's rivers can be viewed as one solar energy system, certainly the world's oceans are another. Indeed, Bryn Beorse, a scientist and engineer, reminds us in this section's fifth article, the conversion of ocean thermal energy into electricity comprises the world's largest solar energy system. Ocean thermal energy conversion (OTEC) uses warm sea water to heat liquid ammonia into gas, which in turn drives a turbine. The idea works, Beorse tells us, but it needs a commitment from the Federal Government before it can make any contribution at all to the nation's energy supply.

The final two articles in this section examine two sources of energy that owe little to the sun's rays: geothermal energy, from the earth's hot interior, and tidal power, a product of the gravitational forces of the sun and the moon. The French and the Soviets already use tidal power to produce small quantities of electricity. However, geothermal energy will probably play a larger role than tidal power in America's energy future.

George F. D. Duff, a mathematician who has been studying tidal power for more than a decade, explains why it is difficult to harness the power of the tides. He believes that eventually 3 percent of the earth's tidal power could be put to use generating electricity.

Geothermal energy seems a more attractive energy source, as Peter Britton, writer on scientific subjects, points out in this section's final article, "Geothermal Goes East." Iceland already gets much of its energy from wells drilled into reservoirs of superheated water and steam. It has long been thought that the western United States could follow Iceland's example. Now, it seems, geothermal energy is also available in useful quantities in the Northeast. Unfortunately, geothermal energy is neither always easy to find nor, once found, put to work.

SMALL SUCCESSES IN SOLAR ENERGY[1]

Last May [1979] the White House unveiled a showcase solar energy installation designed to supply 44 percent of the

[1] Excerpted from the article "Despite U.S. Nudges, Solar Energy Moves Slowly," by Anthony J. Parisi, staff reporter. New York *Times.* p 1+. D. 8, '79. © 1979 by The New York Times Company. Reprinted by permission.

hot water used in the West Wing. Although the system cost more than $40,000, the solar lobby was ecstatic because the engineers had calculated that if such equipment qualified for tax credits, it would pay for itself in less than seven years.

"I hope we can set an example for the rest of the nation," President Carter proclaimed.

The Energy Policy and Conservation Act of 1975 had required the President to develop and carry out a plan to reduce energy consumption in the Government's 400,000 buildings by, among other means, switching to solar energy.

Four years later there is no plan. And while the White House hot-water unit is said to be working well, there is little use of solar energy in other federal buildings.

Amid the greater frustration of the energy program, solar energy has so far had just such small success. Many energy specialists contend, however, that after six turbulent years that began with the Arab oil embargo followed by the loss of Iranian oil, solar energy is ready to start sharing the burden borne by oil, gas, coal and the atom.

Today, many forms of renewable energy owe their existence directly or indirectly to the sun, from the firewood burned in New England to the manure converted to gas in the Plains States to the hydroelectric power generated in the Northwest. But collectively, they satisfy about 6 percent of the national energy budget, or half again as much as nuclear power. President Carter has set 20 percent as a goal for the year 2000, and solar advocates believe that much, and perhaps more, is attainable.

"We are not running out of energy," said Denis Hayes, director of the Solar Energy Research Institute, the Government's newest national laboratory. "We are running out of cheap oil and gas."

After many false starts, the solar community is retrenching. Many proponents are playing down such spectacular ideas as satellite solar stations, and sometimes even rooftop panels, and emphasizing instead more mundane approaches that are easier to sell, such as grain alcohol for cars or greenhouses and other "passive" solar devices for homes. Accordingly, they are stressing economics, as well as environmental considerations, in their stepped-up campaign.

Society can hope to restrain the rising cost of energy, solar advocates argue, only by shifting from finite fossil and fissionable fuels to renewable energy derived from the sun. They concede that the cost of solar-replenished energy will increase as labor, materials and capital become more expensive, but they argue that such energy will at least escape the accelerating cost of recovering depletable forms of energy once all the easily accessible deposits are consumed. . . .

Solar enthusiasts maintain that their way is the cleanest and safest of all—and available starting now, in small doses anyway, not twenty or thirty years hence. To them the term *solar energy* has come to include the panoply of naturally renewable energy sources, including hydropower, wind, ocean currents, firewood and other combustible vegetable matter. All these are stored solar energy, as are conventional fuels, but these stockpiles are replenished in days, not eons.

To most people, though, solar energy still means simply heat from the sun, and it is here that the biggest push has come. It is also where some of the greatest disappointments have occurred.

"We've Grown Fast"

"We've all heard the horror stories," said Sheldon H. Butt, president of the Solar Energy Industries Association, a trade group that represents seven hundred manufacturers and installers of solar equipment. "But for every solar installation that didn't work, thirty have. We've grown, and we've grown fast."

The association's figures show that in 1978 five million square feet of solar panels were sold, as against 137,000 in 1974. Lately, though, sales have slowed, and the manufacturers blame the Government's solar tax credits, which Congress passed last year [1978].

Although the credits allow residential buyers of solar equipment to reduce their income taxes by up to $2,200 (30 percent of the first $2,000 and 20 percent of the next $8,000), the program was delayed a year and a half while Congress debated the rest of President Carter's energy package. Many potential customers postponed purchases pending the outcome, and the manufacturers complain that momentum was lost at a formative stage of their infant business.

Others, however, say another, more fundamental, factor seems to have contributed to the slowdown. Despite the tax credits, rooftop panels are still too costly for the vast majority of homeowners. To date, only solar hot-water heaters have proved feasible in most locales. Unlike systems for heating the houses themselves, these operate year-round and therefore pay for themselves in half the time.

In California, both San Diego and Santa Barbara counties now require solar hot-water heaters on all new electrically heated homes. The units pay off even outside the Sun Belt because the saving depends not only on the amount of sunshine but also on the cost of electricity. In the not-so-sunny New York area, where rates are up to twice the national average, solar hot-water heaters can be a good investment, the experts say.

But systems designed for space heating rarely survive the economic scrutiny of hard-nosed buyers, even with the tax credits. Aware of this, many proponents of solar heating are having second thoughts. They now regret the early rush into so-called active systems, the familiar rooftop installations replete with pumps, valves, switches and other potentially cranky gadgets, all of which often add up to an eyesore. In the haste to embrace solar heating, these supporters say, people seem to have overlooked the myriad improvements that are possible simply by designing houses to accept the sun's energy passively.

Passive solar features include elements as simple as picture windows or as elaborate as solariums. The general idea is to open an otherwise energy-tight house to the environment, but selectively, letting solar heat enter in winter and escape in summer. Although the contribution of each passive feature may be minor, together they can make an appreciable contribution. And they often cost nothing.

"You have to orient the building some way," said Fred

Dubin, a partner in the New York architectural firm of Dubin, Bloome Associates, which specializes in energy-conscious design. "You might as well orient it to get a large southern exposure if you can."

Solar architects urge builders to minimize glazing and maximize insulation on the north side, which is always an energy loser, and to place plenty of glass on the south side, which in the northern hemisphere is usually an energy gainer. To block solar rays in summer, when the sun climbs high, they suggest overhangs or other shading elements. They encourage homeowners to plant evergreens on the north side to break wintry winds and deciduous trees on the south side to provide still more summer shade. . . .

Passive solar techniques can slash fuel consumption by 40 to 50 percent, according to . . . Dubin, the New York architect. That, he said, would come on top of the 35 to 50 percent saving that is possible by specifying the right amount of insulation and taking other steps to retain heat. Together, . . . Dubin estimated, efficiency improvements and passive solar considerations should add no more than 10 percent to the cost of a new house.

Those in the solar movement who have come to favor passive designs see them not necessarily as an alternative to active systems but as a sensible step in the solar transition.

Picking Economical Method

"It's silly to install a piece of solar equipment on a house with broken windows," said Henry C. Kelly, the Solar Energy Research Institute's director of policy analysis. "We should prioritize our investments. If it's more economical to conserve, then conserve. If passive is more economical than active, then use passive. If active is more economical than electricity, then use active."

Dr. Kelly added, however, that the costs were different for the nation as a whole than they appear to individual consumers. When buyers compare solar power with what they would have to pay for electricity, he explained, they are using electricity prices that reflect the present cost of power, not its replacement cost, which these days is considerably higher.

"The national objective must look at the marginal cost, not the average cost," he asserted. "It must also include the externalities," he added, referring to the social costs of more air pollution and waste disposal. Including those factors, the cost of electricity looks more like 9 cents a kilowatt-hour than the 5 cents or so that consumers pay on average nationwide, he said.

Although the Government's tax credits are designed to offset some of this imbalance, solar enthusiasts maintain that still more incentives are justifiable. With them, they say, even active systems would soon become commonplace.

Meanwhile, solar manufacturers are developing fresh approaches that could bring the cost of active systems way down: plastic collectors that are extruded, cut off by the yard and nailed to the roof; cylindrical systems made with ordinary fluorescent tubes, and solar collectors built into the roof while the house is going up.

"The sun has become a serious alternative source of energy," says Modesto A. Maidique, a solar expert at the Harvard Business School. "The kind of relatively low technology needed for a 20 percent contribution by the end of the century is basically already here. What stands in the way are institutional barriers, plus a lot of bad habits."

ENERGY FROM BIOMASS[2]

Most of us have forgotten or never realized how much this country depended on energy from biomass during its first hundred years. The earliest uses of such energy in this country involved wood for heating and cooking, grass and hay for ani-

[2] Reprint of article entitled "Energy from Biomass—Coming Full Circle," by H. Leon Breckenridge, manager, Economic Analysis and Biomass Energy Systems, EG&G Idaho Inc. *USA Today*. 107:48–50. S. '78. Reprinted by permission. Copyright © 1978 by the Society for the Advancement of Education.

mal power, and, in certain areas, conversion of wood into charcoal or coke for industrial use. As late as 1850, biomass provided for over 70 percent of the US primary energy requirements.

During the nineteenth century, a major (and probably an unfortunate) consumption of wood biomass produced charcoal and coke for use in making iron, steel, and copper. This practice denuded hundreds of square miles of forests in many locations. Wood used to be almost the only source of energy for heating and cooking in this country. This use of wood continued through most of the nineteenth century. In forested areas, wood still provided all or the majority of the heat for many homes.

Oxen and horses are both forms of power which are derived from biomass. Lancaster County, Pennsylvania, is an area where this form of energy from biomass is still utilized extensively for farming. It was not until 1947 that the number of tractors exceeded the number of horses and mules on farms in this country.

Around the turn of the century, the use of energy from biomass decreased as oil and coal became the predominate sources of energy. During this era, few locations produced what was called "town gas" manufactured from wood and coal. This gas had a low energy content, but could be used satisfactorily for cooking and heating. By the 1920s, there were over 20,000 coal gasifiers producing gas in the United States. This process for producing "town gas" was pyrolysis-gasification, which can convert any organic material into a clean-burning fuel.

Another old technology which is still in significant use today is the production of methane gas at sewage treatment plants. Methane is the primary ingredient of natural gas and can be substituted on a one-for-one basis. This process is the anaerobic [without air] digestion of organic material by specific microorganisms. Nature has done this since before the coming of man. There is strong evidence that the first living organisms on Earth were anaerobic bacteria. It is due to the effort of these little bugs for hundreds of millions of years that we are able to take advantage of the enormous quantities of natural gas and oil that we do today. The Indians in America were aware of this natural process. This digestion was, and is, common in swamps, where the decaying materials produce bubbles of gas. This gas contains methane and can be readily burned.

Perhaps the oldest method in continuous use for conversion of biomass into useful energy is fermentation. When suitable bacteria are used, grains, fruits, and many other kinds of biomass can be converted into a useful fuel. However, most cultures have chosen to use it as a beverage. Many call it beer, wine, or spirits, all of which contain ethanol. If different types of bacteria are used with other feedstocks, usually a form of wood, then the product is methanol. Both of these alcohols are exceptionally clean-burning fuels. [For more on methane see "What's Wrong With Gas?" by Eugene Luntey, in Section IV, above; on ethanol, "Alternative Fuels for Cars," by L. R. Brown and others, in Section V, above.—Ed.]

Technologies for Conversion of Biomass

The pulp and paper industry has always been a large user of biomass. During the 1890s and early in this century, some plants in the United States were energy self-sufficient, producing all of their electricity from wood waste. Today, a typical pulp and paper mill will generate half of its power by burning biomass to produce steam, which is then used to turn steam turbine electrical generators.

The earliest pulp and paper plants burned hog fuel—waste wood products such as bark, trimmings, and other unusable parts of the tree. The boilers used to burn the hog fuel were very large and costly to build. Later, the amount of available hog fuel decreased due to improved utilization of the tree. As a result, black liquor boilers were added to provide electricity and process steam. Black liquor is a waste byproduct from the conversion of wood chips into paper pulp.

Another method of converting biomass into a useful fuel is pyrolysis-gasification. This process produces a gas that can be utilized in steam boilers, many gas-fired ovens, and large internal combustion engines. Variations in the design and operation of the equipment will result in other combustible products, such as a thick oil or charcoal. Depending on the specific design, a low-, medium-, or high-energy content gas can be produced. Some of each of these systems are in operation today.

Pyrolysis-gasification may be described as a combination of two well-known and extensively used processes from the wood and coal industries. The char gasification reactions are chemically the same as those used for the production of gases with low-energy contents, known as "coal gas," "manufactured gas," or "water gas." The last term was coined to identify one of the major reactions, which is the steam (water)-carbon reaction. The second process, destructive distillation, is used in the production of coke from certain types of coal and is also used in the production of charcoal from wood. All chemical processes that occur in pyrolysis-gasification of wood also occur in a wood-burning fireplace in a home.

Anaerobic digestion of biomass consists of two distinctly different stages. In the first stage, bacteria act upon the complex organics in the biomass to change them into simple, soluble organic material called volatile acids. The second stage involves a different group of bacteria, which converts the volatile acids by a process called fermentation into a mixture of methane gas and carbon dioxide gas. There are a number of different strains of bacteria in each of these stages. As a result, some methane and considerable carbon dioxide is produced in the initial stage. By selecting and maintaining specific bacteria for the final stage of fermentation, pure methane could also be produced. The chemistry of this process is quite complex and not well-understood. It is very difficult to obtain optimum yields, while, at the other extreme, it is almost impossible not to get some burnable gas from the most crude experiment, which anyone can perform in the backyard.

To maintain continuous production at reasonably high levels of efficiency, a proper balance between the two groups of bacteria is required. To maintain this balance, five environmental parameters must be controlled—temperature, anaerobiosis, acidity, nutrients, and toxicity of input. The two major groups of microorganisms require a different range of temperatures for optimum production. Maintenance of anaerobiosis [life without air] in the digester is essential. Even small amounts of air can almost totally disrupt the production of methane. Acidity must be maintained over a fairly narrow range. This is a difficult task since, in the first stage, the bacteria are producing acid and, in the second stage, different bacteria are converting acid to gas. Another requirement is provision of adequate nutrients for the various bacteria. The rate of methane production is closely related to the ratio between carbon and nitrogen in the digester feedstock.

For successful production of gas from biomass, the fifth environmental parameter, that of toxicity, must be maintained at a low level. Compared to the other four variables, very little is understood about how much and what combinations of unhealthy trace substances in the feedstock will produce a toxic level for the different bacteria in the digester. Fortunately, with most biomass, other than municipal garbage and sewage sludge, toxicity is not a problem and some gas will be produced over a wide range of each of the other parameters. It is only when optimization is attempted that high technology is required. For example, in India, farmers and other rural families have been using biomass digesters for over forty years. These have been made at the local level with materials that are readily available. It was estimated that, in 1970, well over 250,000 of these units were in use in India.

Another country which has gone to biomass fuel in a big way is China. Only a few digesters were in operation before 1970. By 1975, over 200,000 units were in use just in the province of Szechwan. By 1977, there were an estimated 4.3 million digesters in use in China.

Sources of Biomass

All agriculture crops that are grown for food which will be used for human consumption produce many times as much waste as they do an edible portion. Most of the crops grown for livestock feed also produce large quantities of waste material. Significant quantities of organic wastes result from the processing of fruits, vegetables, cereal grains, and other foods.

Not all of the wastes left in the field can or should be considered as available for conversion into energy. To do so would deplete soil nutrients and cause subsequent loss of crop yield. There are methods of conversion of these wastes into a burnable fuel that allow most of the essential nutrients to be returned to the soil. Under these conditions, 75 percent or more of the crop (biomass) wastes would be usable for conversion to other energy forms.

A different form of agriculture is marine farming or aquaculture for the production of very large quantities of biomass. One method which is under development off the California coast is giant kelp farming. Yields are very high—35–60 dry tons per acre per year. The kelp will be harvested and converted to a pipeline-quality gas by anaerobic digestion. A similar project is being considered in Florida by growing water hyacinths in canals. The yield per acre per year is also very high—about 60 dry tons. An additional benefit may be the reduction of the $15 million annual cost to the state of Florida for water hyacinth control in existing waterways.

Other water-based farming methods proposed for production of biomass include algae grown in the ocean, in cooling ponds of electrical power generating plants, or in converted dry lake beds. The algae would be converted to useful energy either by anaerobic digestion or fermentation. Until the Solid Waste Disposal Act was passed in 1965, mixed municipal solid wastes (MMSW) were considered a nuisance and were disposed of in the cheapest way possible. Much of the industrial and most of the commercial burnable solid wastes end up in the municipal solid waste stream. Later, the act was amended as the Resource Recovery Act of 1970. Since that time, various technologies have been developed or improved which can recover the biomass (organic) portion of the MMSW and convert it into useful energy (fuel).

Approximately, 40 percent of MMSW is paper and other forest products. In addition, there is another 40 percent that consists of yard wastes, food wastes, and other organic materials. On a national basis, 80 percent of MMSW is one form or another of biomass and can be converted to energy with one or more technologies.

Recent environmental awareness has resulted in significantly improved utilization of the "total tree" by the forest products industry. Many articles in forest product industry literature emphasize total tree utilization. Some of these suggest that 80–90 percent conversion of the tree is possible. This high utilization refers only to that portion of the tree normally harvested and delivered to the mill.

Work at the University of Maine has been performed to do a complete biomass inventory of the forest. Using the total biomass as the basis, only about 50 percent of it is delivered to the mill and utilized under so-called total tree utilization. The remaining 50 percent is a very large source of biomass that could be converted to useful energy.

The oil embargo and subsequent drastic increases in the cost of energy have resulted in increased utilization of biomass wastes at pulp and paper mills. Some authorities predict that, if or when a new grass roots mill is built, it would be energy independent from outside sources and utilize 100 percent of all the organic (biomass) material delivered to the plant site. Therefore, any estimates of available biomass should exclude those forest materials that can be utilized for energy at the mill.

Projected Quantities of Biomass

Most biomass sources are not considered a waste and disposal is not a factor. For this reason, collection and transportation of existing biomass residues is the primary economic consideration. The table below gives an estimate of biomass wastes generated in the United States, but does not consider how much of it is economical to recover.

The approximate nine quads (one quadrillion or 10^{15} BTU) of net gas potential for 1971 was equal to two times the amount of gas used for heating and hot water for all residences in the United States in 1970. In fact, this amount of gas could have replaced all the natural gas, oil, and electricity used for household heating in 1970!

In addition to the existing wastes, a very large potential exists for growing biomass on idle crop land. According to the Department of Agriculture, 10 percent of class I through class IV lands (lands that can be cultivated and are not arid) were idle in 1974. Biomass grown on these 20 million acres could have produced 3.9 quads of energy.

An example of how drastic a change in thinking has occurred is a study by the Governor's Energy Advisory Council for the State of Texas made between 1973 and 1975. Texas oil production in 1976 was about six quads. They estimate that, by the year 2000, there will be 0.67 quads produced from biomass and three quads from other solar methods. The largest oil-producing state is predicting that, by the end of this century, the amount of energy its citizens will be using that is derived from direct sunlight, in one form or another, will be

Estimates of Organic Wastes Generated in the United States, 1971 and 1980

Waste Type	1971	1980
	(Millions of tons per year)	
Agricultural Crops & Food Wastes[a]	390	390
Manure	200	266
Urban Refuse	129	222
Logging & Wood-Manufacturing Residues	55	59
Miscellaneous Organic Wastes	50	60
Industrial Wastes[b]	44	50
Municipal Sewage Solids	12	14
Total	880	1061
Net Oil Potential, (million barrels)[c]	1098	1330
Net Gas Potential, (billion cubic feet)[d]	8800	10600

[a] Assuming 70% dry organic solids in major crop-waste solids.
[b] Based on 110,000,000 tons/yr. of industrial wastes in 1971.
[c] Based on an oil yield of 1.25 bbl/ton of dry organic waste.
[d] Based on a methane yield of 5 CF/lb. dry organic waste.

equal to over one half their current oil production. The old expression "make hay while the sun shines" will take on new meaning if that ever comes to pass.

HARNESSING THE WIND[3]

Wind is generated by our sun, that thermonuclear marvel that is close enough to Earth to pour life-giving rays of energy upon it but far enough away to prevent our big blue marble from roasting to a crisp. Variations in the sun's heating of the Earth's surface and atmosphere due to differing cloud, climatic, and landscape conditions cause air to move. This movement has tremendous power—equal, according to estimates, to several times the yearly total of energy used in the United States.

But can the power of the wind be harnessed to relieve the current energy crisis? The answer is still uncertain. Wind power is attractive because, unlike fossil fuels, its supply is constantly renewed and the process of capturing it is virtually nonpolluting. Unfortunately, however, the same variations that guarantee that wind will blow on Earth as long as the sun exists also make the wind an unpredictable, undependable source of energy—at least as a single substitute for the powerful and reliable but fast-disappearing fossil fuels upon which industrial society desperately depends.

Fossil fuels are a highly concentrated form of energy that can be stockpiled to provide continuous power. But wind is a dispersed form of energy that is difficult to store and seldom blows with consistent force. It varies widely from place to place (even at sites only a few hundred yards apart); it changes direction frequently; and somtimes it doesn't blow at all.

These characteristics add up to one major obstacle to pro-

[3] Reprint of article by Lee Stephenson, free-lance writer, former editor of *Environmental Action*. *National Parks & Conservation Magazine*. 52:10-15. My. '78. Reprinted by permission from *National Parks & Conservation Magazine*, May 1978. Copyright © 1978 by National Parks & Conservation Association.

ducing power from the wind for mass consumption: high cost. Storage or the need to maintain backup systems seem to make most wind power installations more expensive to users than conventionally produced electricity. Such comparisons are misleading, however, because mass production cost of wind generators is still undetermined and because the many federal and state subsidies given to the fossil fuel and nuclear power industries are overlooked in most analyses.

In spite of these problems, governments, private companies, and hundreds of inventors are now attempting to find less expensive ways to harness the wind. The basic technology has changed little since windmills were first used in Persia more than two thousand years ago. Blades or other objects are used to catch the wind, turning a shaft that powers a grinding wheel, a pump, or an electric generator. Gears are often used to increase the rotational speed of the shaft. Some of the earliest windmills operated on a vertical axis to capture wind blowing from any direction, but most have been horizontal-axis machines that must be turned, either automatically or by hand, to face the wind as its direction changes.

Officials of the US government's Department of Energy (DOE), which now has the largest known wind-power development program in the world, believe that the best way to ensure that wind power will make a significant contribution to the nation's energy needs is to make it economically competitive with conventional power sources. While experimenting with small systems, DOE sees large wind turbines as offering the greatest potential contribution. One of its major goals is to make it economically attractive for electric utilities to purchase large machines and connect them to existing power systems. Others, including the growing number of small wind turbine manufacturers, claim that small wind generators purchased by individual households, office buildings, or neighborhoods are the most promising application.

All these approaches have been tried and abandoned in the past; but contemporary wind-power enthusiasts hope that modern, light-weight metal alloys and new manufacturing techniques—as well as the constantly rising cost of electricity generated by fossil fuel and nuclear power will spell success for wind generators. Windmill technology was used in the Middle East in the eleventh century and in Europe in the thirteenth century, and Dutch settlers brought the windmill to America in the mid-1700s. More than six million small wind machines were built and sold here between 1850 and 1950. These machines were primarily water-pumping units, but several hundred thousand electricity-producing machines were sold after 1900. The demise of this early domestic windmill industry came after the formation of the Federal Government's Rural Electrification Administration in the 1930s, which brought power lines to much of the rural United States for the first time. Thousands of these early windmills, some of which remain functional, can still be seen on farms in many areas of the nation.

Large-scale generation of electricity also has a long history. Around 1900 the Danish government produced the first of a generation of giant experimental wind turbines. Mounted on an eighty-foot tower, the Danish machine had four blades with a diameter of seventy-five feet and produced from 5 to 25 kilowatts of electricity. (A kilowatt is equal to one thousand watts; a watt is a unit of power equal to 1/746 horsepower.)

The Russian government followed in 1931 with the construction of a 100-kilowatt wind generator that was connected by electric lines to a conventional power plant twenty miles away. The largest wind machine of this period was designed by a US engineer, Palmer Putnam, and was constructed in 1941 near Rutland, Vermont. Connected to the Central Vermont Public Service Company's electric system, Putnam's generator produced its maximum power of 1,250 kilowatts in a thirty-mile-per-hour wind. The project was abandoned in 1945 because replacement parts were unavailable during and after World War II and because of doubts that the generator could ever be economically competitive.

The British, French, Danes, and Germans also built large-scale experimental wind machines between 1940 and the early 1960s. All these programs, however, were scrapped by the late 1960s because none of the generators could produce power at a cost competitive with then-inexpensive fossil fuels. Now a number of nations, including Germany, Sweden, and Denmark, have begun new wind research programs.

After a review of the past attempts, US scientists designed a two-blade, horizontal-axis, 100-kilowatt test generator that was constructed in 1975 at a government research center near Sandusky, Ohio. Based on those tests, three similar but more powerful wind generators are being installed in three existing utility power grids. The first installation, at Clayton, New Mexico, has been completed; the local electric utility will soon begin using the wind turbine. The other sites are Culebra Island, off Puerto Rico, which is scheduled to begin operation later this year [1978], and Block Island off Rhode Island, scheduled to be completed in 1979. Each of these 200-kilowatt generators will produce enough electricity to supply about sixty homes when the wind is blowing at least nineteen miles per hour. Also in 1979 DOE plans to install a much larger machine at Boone, North Carolina—a 1,500-kilowatt (1.5 megawatt) generator with two-hundred-foot-diameter blades that could supply electricity to five hundred homes in winds of twenty-two miles per hour or more.

One island community off the coast of Massachusetts has bought and installed its own 200-kilowatt wind generator from a New York company. Cut off from electricity lines on the mainland, the citizens of Gosnold are tired of paying high prices to ship in oil to power diesel generators, and they hope to get cheaper electricity from their wind turbine when it begins operation in mid-1978. The diesel generators will supply backup power during times of insufficient wind.

An advanced concept in the DOE research program that some think is as promising as these horizontal-axis units is a modern version of the first vertical-axis windmills in Persia. Called the Darrieus rotor, it has the advantage of catching wind blowing from any angle.

From an environmental perspective, wind energy is especially attractive. Although pollution is always released in the processes that use energy, wind generators themselves create virtually no pollution. One might question the esthetic impact of having large numbers of windmills spread over the landscape, and some concern has been voiced that birds may be killed in collisions with wind machines. Also, significant pollution may be created in the manufacture of wind generator equipment.

Most of the work of the DOE wind program, including studies to determine development priorities, has been conducted by aerospace corporations under contract to the department. This arrangement has led to criticism of DOE because most of these companies are experienced only in complex, centralized technologies. Until recently only a minor portion of DOE's wind power budget was devoted to small wind machines—those that produce less than about 100 kilowatts. The reason for this early bias, according to Donald Teague, a program manager for the DOE wind project was that government decisionmakers believed that large-scale field tests were needed to demonstrate the potential of wind power. Teague says it will be difficult to make small machines competitive enough to gain widespread public use. He points to the $8 million in funds earmarked for small machine development and testing in 1978 (25 percent of the wind program's total budget) as the beginning of a major effort to find out if this widespread use can be achieved.

Ironically, sales of small windmills, manufactured by a host of new companies, are booming. Only one or two manufacturers survived the 1960s, but today at least seventeen new US firms, primarily very small companies, are selling a variety of small water-pumping or electricity-generating wind machines. The DOE has estimated that 2,500 water-pumping and 1,100 electricity-generating wind machines were manufactured in the United States in 1976. The electric models cost from $2,000 to $15,000 or more to purchase and install; water-pumping models run from a few hundred to several thousand dollars.

A number of the small manufacturers, as well as some distributing companies and consumers, have joined forces in the American Wind Energy Association, based in Bristol, Indiana, to promote their goals to the Federal Government and to provide information to potential buyers. Rick Katzenberg, the current president of the group as well as the president of Natural Power, Inc., a New Hampshire distributing company, says the group supports the work of the Federal Government on large machines. He adds, however, that the association's members believe that small wind machines are equally deserving of federal attention because they may offer the opportunity to reduce the need for expensive regional power systems and give those people who want it the means to produce their own power. "We've pushed for more of a balanced approach," Katzenberg said, "and the gap is narrowing significantly."

The best evidence for DOE's increasing support for small wind machines, according to Katzenberg, is the establishment of the Rocky Flats federal testing center at Golden, Colorado. The DOE-funded center has purchased a number of the available small-scale wind machines and will conduct standardized tests on each. Plans call for the testing of all available US-manufactured wind units and for publishing of the test results.

Rocky Flats is also the project manager for the development of nine new systems by private industry. These systems include three 1-kilowatt, four 8-kilowatt, and two 40-kilowatt units.

Storage systems remain a problem for both small- and large-scale wind generators. Many home or farm wind systems have used batteries to store power, but they are very expensive and must operate on direct current power, making the systems incompatible with most modern appliances. Other potential storage systems are being studied, but so far costs are very high and many technical problems still exist.

One unique "storage system" now being used in several locations is a device called a synchronous inverter, which permits the owner of a wind generator to feed surplus alternating current electricity into existing utility power lines. Any power supplied to the utility can show up as a credit on the wind generator owner's electric meter. This connection also allows the individual to buy power from the utility on windless days.

Although this system works well, it does create some new problems. If a large number of wind generators were connected to a utility grid, the power company would sell much less power on windy days but still would need enough generating capacity to serve all the customers on days when there is little or no wind. Unless a special rate structure were used for the wind generator owners, other customers would pay more than their share of the cost to build generating plants.

A recent test case on this issue resulted in a victory for wind power enthusiasts. The New York State Public Service Commission ruled in 1977 that Consolidated Edison must allow a Manhattan wind generator owner to sell excess power to the utility and buy power from it during windless periods. The company was given permission to charge the wind generator owner a higher rate to reflect special costs.

The least expensive arrangement for small wind generators may be no storage at all and the use of oil-powered backup generators for windless times. Also, it may be most cost effective to use large wind generators without storage to augment the capabilities of existing electric utility plants. Or, if large generators were spread over a big enough area in a power grid, one location's windy times might compensate for another's windless ones.

DOE's Donald Teague points out that it would take up to ten thousand large wind generators to make a significant contribution to the US supply of electricity. The agency's current cost estimates predict that 1,500-kilowatt generators produced in that quantity today would cost about $860 per kilowatt—about $350 per kilowatt more than a new fossil fuel plant completed today and about $200 per kilowatt more than a new nuclear plant. The estimated cost of power per kilowatt hour from these wind machines is about six cents when distributed through power lines. This is about two cents more than the current [1978] national average. [Figures for residential costs in 1979 in the Northeast ran in the neighborhood of 8.5 cents per kilowatt hour.—Ed.] Costs for smaller wind generators now on the market are higher; a preliminary study for DOE estimates a minimum cost of $1,500 for each kilowatt of capacity.

Naturally, as the price of oil, natural gas, coal, and enriched uranium increase, wind machines will become more competitive. In addition, new studies that reflect accurate mass production costs for wind machines and include costs of special subsidies given traditional forms of energy may show wind power to be more attractive.

Trying to estimate the percentage of our energy use that wind power can provide in the year 2000 or at any point in the future is sheer guesswork. Teague believes that most of these questions can be resolved in the next five to ten years.

Some people say the answers could come sooner if the federal wind program had more money. Even though the United States probably has the largest wind research program in the world, wind and other alternative energy forms are still small projects compared to nuclear power development in the federal budget. In fiscal year 1978, wind power comprises $37 million, slightly more than 2.5 percent of DOE's $1,473 million budget for research, development, and demonstration programs. Solar energy is allotted $374 million, some 25.4 percent of the budget; but nuclear energy is granted $1,062 million, 72.1 percent.

A legitimate question can be raised as to how fast such a program can grow without wasting money. Teague says he has mixed emotions on this subject but believes the program could probably advance faster with more funds, particularly for personnel. The entire federal project is now managed by Teague and three others.

Wind energy alone will probably never replace fossil fuels, but it may be an important means to conserve those dwindling supplies. And it is possible that other alternative technologies such as solar heating, photovoltaics, bioconversion, and conservation could be used in combination with the wind to offset one another's weaknesses.

Experts generally agree that wind power is much closer to

being economically feasible than most other alternative energy technologies. If costs are reduced to a competitive level, wind is likely to play a major role in supplying our future energy needs. Even if costs remain higher, society may choose to utilize wind energy widely to benefit from its much less environmentally destructive generating process.

A FUTURE FOR HYDROPOWER[4]

Supplying energy for our future needs is a subject of current concern to nearly all Americans. While we still think in terms of traditional large-scale technology—such as that embodied in nuclear and fossil fuel power plants—to fulfill these needs we have also begun to heavily explore the small-scale and alternative possibilities. The potential of combinations of solar, wind, methane and geothermal energy, and of conservation practices to fulfill our demand for energy with minimal environmental impact and on a scale adaptable to local needs and to local control are now being included in nearly all discussions of future energy supplies—although, of course, this does not mean that these alternative sources are truly being given the consideration they deserve.

Missing from nearly all of these discussions of "appropriate technology" is hydroelectric power. Although hydroelectric power is a clean and renewable energy source, it is perhaps not surprising that we fail to include hydropower in our alternatives for the future. Today the term *hydroelectric power* brings to mind the image of towering concrete dams—an image accompanied by negative environmental and economic connotations.

Yet flowing or dammed water was one of the first power sources for generating electricity, and it did so on a modest scale and with minimal environmental impact. In particular,

[4] Reprint of article by George S. Erskine, an electrical engineer. *Environment.* 20:32-8. Mr. '78. Copyright © 1978 Helen Dwight Reid Educational Foundation. Reprinted by permission of the Foundation and the Scientists' Institute for Public Information.

the waterwheel, used since antiquity for grinding and pumping, was converted into a system to drive electrical generators. The waterwheel was not particularly efficient in this service and modern turbogenerators which could withstand high
water pressure replaced the old-time waterwheel. Huge dams
were built to create this pressure and to supply the massive
amounts of energy demanded by an expanding society and
technology. Today, however, faced with soaring costs for fuel
to power steam generators—and at the same time realizing
the economic and environmental impracticability of large
dams—engineers and others are earnestly investigating modern ways to tap the renewable energy in moving water without building gigantic dams.

The best-studied possibilities for this small-scale type of
hydroelectric power generation have involved the utilization
of the power potential of existing dams. The US Army Corps
of Engineers estimates that current hydroelectric capacity
could be nearly doubled if these dams were retrofitted for
turbines and generators and fully utilized. Less studied are
nondam methods of producing small-scale hydroelectric
power. Several possible ways for utilizing this free-flowing
power without creating reservoirs will be described in this
article. No estimates have been made of the amount of power
that could be potentially derived from this type of hydropower.

Potential at Existing Dams

A recent study by the Corps of Engineers, entitled "Estimate of National Hydroelectric Power Potential at Existing
Dams" [submitted to the President, July 1977], assesses the
potential for increasing hydroelectric capacity by retrofitting
existing dams. The study was the result of a ninety-day assessment of the power potential at the 49,500 dams in the United
States designed to provide background information for President Carter's energy proposals.

The Corps study considers four types of possibilities for
utilizing existing dams: (1) renovating and improving existing

hydroelectric facilities; (2) installing additional turbines and electrical generators at existing hydroelectric dams; (3) installing generating equipment at large dams that were built for purposes other than for producing electricity and do not now have generating capacity (retrofitting); and (4) retrofitting small dams—those with less than 5,000 kilowatt potential—for very small-scale, local power supplements.

Central to the potential for new ways to use existing dams—and also the nondam hydropower sources to be described later—is the bulb turbine. The bulb turbine is a modern adaptation of the old-fashioned waterwheel-generator combination. This turbine, like all water turbines, works on the same basic principle as the waterwheel: its blades are turned by the pressure of flowing water. It is named for the bulb-shaped housing which protects the generator. The turbine, which drives the generator, is external to the bulb housing, but both sit within whatever aqueduct or pipe is carrying the water.

The bulb turbine is an extremely versatile device; it can be used in conjunction with different types of dams and other water controlling systems—systems that create pressure to turn the turbine's blades. The bulb turbine, because its generator is protected, can even generate electricity by being directly submerged into fast moving streams.

The tube turbine can also be used in some of the instances to be discussed but is less flexible than the bulb turbine. Although both turbines work in the same general manner, the tube turbine's generator is not encased in a submersible housing. The turbine must be contained within the water-carrying pipe or aqueduct, but the generator must be placed outside of the water stream, a less adaptable configuration for some uses. For example, bulb turbines can be placed directly in straight piping; tube turbines require offset piping alignment in order to place the generator external to the water-carrying pipe.

Retrofitting existing dams involves installing modern bulb turbine, tube or crossflow turbine generators to use the water pressure developed at the outlet works. The arrangement of the generators depends upon the structural features of the dam. Dams with outlet works offer the most cost-effective opportunity for retrofitting. Outlet works are designed to release water from the dam reservoir in a controlled fashion. An intake structure at the bottom of the reservoir conveys water to a pipe leading to a special outlet on the downstream side of the dam. In some cases it is possible to install a bulb generator in association with this pipe, utilizing the water pressure of the reservoir to generate electricity as the water is discharged. If the volume of water is sufficient, bifurcators can divide the flowing water into more than one stream, permitting the installation of more than one turbine and generator.

If the dam is not equipped with outlet works, water pressure can be controlled and directed by a siphon-type pipe, which literally siphons water from the reservoir over the lip of the dam and down past a turbogenerator on the downstream side of the dam wall. Siphon pipes can often be installed over the crest of an existing dam or spillway with little change to the dam structure.

Estimated Hydroelectric Capacity

The Corps of Engineers study indicates that by upgrading and expanding facilities at existing dams and by retrofitting other US dams, current hydroelectric capacity could be nearly doubled. Current capacity is 57 million kilowatts, with another 8.2 million killowatts to be available at plants under construction. According to the Corps report this capacity could be boosted 54.6 million kilowatts to bring us to a total of 119.8 million kilowatts of hydroelectric capacity; 26.6 million kilowatts, or nearly half of the estimated 54.6 million kilowatt additional capacity that retrofitting and improving existing dams would provide, was predicted to come from retrofitting *small* dams—those with less than 5,000 kilowatt potential capacity. These small dams are particularly suited to the versatile bulb and tube turbine systems. Their adaptability to local use and local control make them a particularly intriguing source of additional hydropower.

Some limitations and competing demands reduce the theoretical capacities estimated in the Corps study. Demand for water for other uses—for maintaining minimum stream flows for fish and wildlife, for drinking and irrigation supplies, and for recreation—could conflict with storing enough water behind dams to permit power generation of this type. Allowing for these competing demands, the study estimates that approximately 9.3 million kilowatts of potential capacity should be eliminated from the national potential of 119.8 million kilowatts.

The poor physical condition of older dams and reservoirs also reduces the theoretical potential capacity of existing dams; many have deteriorated too far to permit renovation for hydropower. Reservoir storage capacity at some older dams has been reduced by continuous siltation—a process which eventually fills in dam reservoirs, rendering them useless. Potential hydroelectric capacity would be reduced by an estimated 2.8 million kilowatts due to these two problems. Serious thought is being given to desilting these dams, thus mitigating this problem. Finally, another 1.4 million kilowatts of potential hydropower are located at dams which are on streams that are dry for a significant period of time. Nevertheless, even subtracting for competing water demands, poor condition of old dams, and low-water dams, the added power potential would be over 40 million kilowatts and the total US hydroelectric potential more than 100 million kilowatts.

Nondam Hydropower

Although national attention is centered on improvement of conventional hydroelectric capacity at US dams, a great potential exists for the installation of bulb-type turbogenerators placed directly in rivers or in the modern equivalent of sluiceways from rivers, and even in aqueducts and irrigation pipes. The potential hydroelectric capacity of these sources has not been estimated; they constitute a hydroelectric power source *additional* to the 119.8 million kilowatts estimated in the previously-cited Corps of Engineers report.

With sufficient year-round stream pressure and volume,

an entire underwater generating system can be placed directly in a river. This installation would consist of an aqueduct or pipe which would conduct the water to the turbine, the bulb turbine and generator, and a tailrace. The intake pipe controls and concentrates the water stream driving the turbine and would be protected by a trash rake to keep debris out of the system.

The length of the intake pipe would depend on the difference in elevation between the inlet structure and the tailrace. The greater the difference in elevation, the more pressure created; the gentler the slope of the incline, the longer a pipe would have to be to maximize the elevation difference. The exact position of the tailrace, which would be located immediately below the turbogenerator, would be such as to capture the maximum amount of energy from the discharge pipe or suction tube. All equipment would be submerged. Electric cables carrying the electricity from the generator could be buried so that the entire system would be out of sight.

A more effective, less cumbersome, and safer approach would be to install the bulb turbine system not directly in the river but rather in a special conduit or pipe called a penstock which could be buried across a horseshoe bend or an "S" curve in the river.

Just as flowing rivers offer an untapped resource for unobtrusive, small-scale electrical generation, various manmade structures which carry water could be equipped with turbogenerators operating on the same principle as the instream and "S" curve installations. . . . Conduits or closed aqueducts, usually owned and managed by irrigation districts or municipalities, could serve this function. Although the need for transporting water usually was the only consideration when these conduits were planned, the volume and velocity of water in the pipes—some of which are many miles in length—offer a potential energy source for the generation of electricity.

These types of pipes were not designed to house turbogenerators, but they can be fairly easily equipped with either the bulb or tube turbine systems. The best place to install the turbine would be at the pipe's lowest point so as to tap the greatest possible pressure. Most pipes of the type discussed here are equipped with energy dissipating valves at this point. To retrofit the pipe for power generation, it usually should not be difficult to install a bifurcator or Y-shaped diversion mechanism to pipe the water around the valve. The turbines would be contained in the new section of pipe.

The configurations of the bulb and tube systems would differ slightly, the tube generator requiring offset piping as previously described because its generator is not submersible. There would be little difference, however, in the amount of energy generated or the installation cost.

Advantages

Despite the availability of hydropower, utilities and industries have not been doing much to increase hydropower utilization. In fact, before the 1973 Arab oil embargo, small hydropower plants were being retired from service at a steady rate.

Yet hydropower has clear economic advantages. The life of a dam power-house is two to three times that of fossil fuel power plants, and retrofitting or expansion can be completed in one to two years, compared to the ten or fifteen years usually required to design and construct fossil fuel plants. The relatively simple equipment involved in hydroelectric power generation results in lower operating and maintenance costs than for other types of plants, and the present trend towards unattended hydroelectric generating stations will further increase this advantage.

The high capital costs required for the construction of large dams and the installation of generating equipment, generally greater than those associated with thermal plants and equipment, have been an economic drawback to hydroelectric development. But where the dam or waterworks already exists, and the costs have been charged against irrigation or domestic uses, the economics are quite different. In these cases, the capital cost outlay per kilowatt of capacity for retrofitting an existing facility for electric generation is often lower than by any other means. In addition, inevitable increases in fuel costs, in investment costs associated with the siting and construction of fossil fuel plants, and in the costs of air and water pollution control make the clean and renewable power source and simple procedures of small-scale hydropower even more attractive.

Conclusion

The additional energy produced by these alternative hydroelectric sources would by no means eliminate the need for other power generation sources. Although the electrical generating capacity along US rivers has increased steadily since 1920, it has not kept pace with the generating capacity of steam power plants. Whereas hydroelectric plants provided about one third of the US generating capacity in 1920, these plants represent only about 13 percent of the US capacity today. Even if hydropower capacity were doubled, it would still be only a fraction of the total energy picture.

But it is time to begin to think of fulfilling our energy needs with fractions and increments that fit together, complementing each other, to create our entire energy picture. We have been schooled in the belief that great quantities of electricity can be effectively and economically generated only at mammoth sites. The environmental, social, and economic costs of much of this large-scale energy production is now making it an unacceptable—indeed impossible—way to meet our total energy needs. Thus, the incremental energy increase that can be supplied by hydropower from small dams could play an important role in combination with other small-scale local energy production such as solar, wind, and methane. Any increase in energy supply from these clean, renewable sources means a reduction in the demands made on our other, nonrenewable and more hazardous, fossil fuel and nuclear power sources.

CAN OTEC BEAT OPEC?[5]

A growing number of people have realized for about a hundred years that our little globe is bubbling with renewable energy—much more than we could ever possibly use. The oceans receive and store heat from the sun. This vast solar collector and reservoir is available to us without the need for staggering new quantities of copper, aluminum, steel, and land area for installations. By using proven technology we could free ourselves of the necessity of importing oil within fifteen years.

Ocean thermal energy conversion, known as OTEC, is relatively simple and by far the largest of our solar-energy systems. That it goes largely unrecognized is hardly surprising; for the ocean itself has been a perennial mystery. Although it covers most of our planet, only now are we even beginning to reach its depths.

In the tropics Father Sun heats the surface of the ocean from 20° C. to 29° C. Toward the polar regions the temperature is but 1° C. to 5° C. Cold water sinks and flows toward the equator at depths of five hundred or more meters. One of these currents travels from the North Atlantic to the Southern Hemisphere.

Historical Forerunners

In 1881 the French physicist d'Arsonval, the American engineer Campbell, and the Italian scientists Dornig and Boggia proposed to have warm surface water heat ammonia, or some other suitable fluid, to its boiling point. The ammonia vapor would run a turbine and then be condensed by cold water pumped up from deeper layers. The same condensed ammonia would again be fed into the evaporator, thus form-

[5] Reprint of article entitled "Ocean Thermal Energy Conversion," by Bryn Beorse, scientist, engineer, and economist at the sea-water conversion laboratory of the University of California. *Humanist.* 39:12–19. Jl. '79. This article first appeared in *The Humanist* July/August 1979 and is reprinted by permission.

ing a "closed cycle." Other physicists frowned, saying: "No power will come out of those small temperature differences." It is hard to believe that to this day physicists and engineers maintain this outdated view.

George Claude, a pupil of the French pioneer d'Arsonval, created a laboratory-sized plant in Paris in 1926 by a different method. He boiled the sea water itself. As anyone living at high altitudes knows, the lower the air pressure, the lower the temperature at which water boils. Claude lowered the pressure in the boiler by pumping the air out. Evaporation from the surface of the water, as steam, ran the turbine. The steam was condensed by colder water pumped from deeper layers. This condensed water was then removed as fresh water. This "open-cycle" plant produced both power and de-salted water.

From Paris, George Claude went to Ougrée in Belgium where he built a pilot plant capable of producing 60 kilowatts on a continuing basis. Then, with his own money, funds from a few friends, and the goodwill of the American Society of Mechanical Engineers, he went to the shore of Cuba. On his third try he was successful. Through eleven days he produced at the rate of 22 kilowatts. The mounting glut of inexpensive oil and natural gas, however, left his technology undeveloped.

In the 1940s, Energie des Mers, formed by both private and government personnel and capital, researched and de-signed an open-cycle plant. This plant was started at Abidjan, West Africa, in 1947–48, and was designed to produce 14 megawatts and desalted water. The cold water pipeline was built; and corrosion and biofouling, or build-up of slime, were carefully recorded for six months. Then the proposed plant was redesigned for the French Antilles; but before it could be built, France surrendered to the nuclear vision and decided to go for nuclear instead of OTEC.

When I was the managing director of a small American firm manufacturing dynamic balancing machines with elec-tronic registry and other hardware, I visited Energie des Mers. The vast potential of energy from the sea impressed me to such an extent that I dropped everything else. I stayed on to study and work. It was sad to watch the French succumb to the siren song of the nuclear age.

In the early 1950s, Dean Everett D. Howe, of the mechanical engineering faculty of the University of California at Berkeley, established a sea-water conversion laboratory. Joining him, we built three open-cycle plants, the last one with a turbine that produced 9 kilowatts in addition to ten thousand gallons of desalted water each day. From these successful pilot plants we gathered important data for future installations.

Ready to Go

The age of sea solar-power is ready to be ushered in. The technology for OTEC is simpler and does not have as many unknowns as nuclear development. The plants are easy to build. The fuel is free. The waste product is fresh water rather than radioactive substances. At present prices, baseload electricity could be produced more cheaply than by nuclear power.

Price and safety are not the only considerations. Dependency upon dwindling oil supplies is leading to international tensions and further discrepancies among the nations. As supplies of oil for conventional wars run short, wars now may well be nuclear.

Far-Reaching Benefits

The late Dr. Emile Benoit, a Harvard-educated economist who served in our State, Defense, and Labor departments, and Dr. David F. Mayer, an OTEC "hardhat" from the University of New Orleans, conceived an emergency energy program. In their notes we find comment in reference to OTEC:

We would close the gaping hole in our national defense posture resulting from reliance on imported fuels derived from depletable sources. We would restore our trade balance. The benign OTEC and similar technologies would make some developing nations self-sufficient in energy. We could readily help others to develop a reasonable energy supply adjusted to their local economies. World stability would be in sight. We would change the United States' international image from exploiter to benevolent distributor of ecologically low-cost energy systems. No program of lesser intensity can promise so much. The economic, political, and military benefits outweigh the costs.

Cost Effective

Bechtel, Lockheed, Linn, Westinghouse, TRW, and Sea Solar-Power are looking at OTEC in realistic terms. They are devoting increasing expenditures to its development and demonstration.

The technologically oriented corporation, TRW, has stated to the United States Energy Department that it can build ocean thermal plants competitive with costs per kilowatt of nuclear plants, at the present price of uranium. A New Orleans shipyard has offered to build OTEC plants for half the building costs of nuclear plants. Lockheed points out that it takes more than 3,000 BTUs (British Thermal Units) to produce 1,000 BTUs from either a coal-fired or oil-fired power plant; and by contrast only 125 BTUs are required for an ocean plant.

The applied physics lab of Johns Hopkins University indicated in 1976 that the cost for OTEC would be about seven hundred dollars per kilowatt. Today it would be somewhat higher. The University of Massachusetts has used the figure eight hundred dollars per kilowatt.

At Carnegie-Mellon in Pittsburgh, Clarence Zener, John Fetkovich, and Abrahim Lavi are in the forefront of OTEC-related studies. Dr. Lavi has provided a useful account of the reasons for the wide range of OTEC cost estimates. Variations do not represent carelessness but the opposite. Cost per square foot of the areas of heat exchanger surfaces, heat transfer coefficients, and the length of cold water pipelines make for variations. There is also the difference in materials. Lockheed, TRW, and Westinghouse have chosen titanium for heat exchange material; it is a costly metal, but it's strong and corrosion-resistant.

Professor William E. Heronemus of the University of Massachusetts, a navy captain involved in shipbuilding in World War II, maintained in 1975 that "the world has never seen another industrial effort so easy to get started and so capable of producing prodigious numbers of high-class products." Lack of response from the government and energy-centered companies led him this past year to remark: "The energy corporations say: 'Don't bother us now. We've made plans to sell all that remains of that oil and gas and that coal, and of that uranium and plutonium.' That's really at the heart of the whole problem."

Worriers and Critics

As a nation haven't we often worried about cost at the wrong time? There is no possibility for exact estimates before we have had OTEC plants operating for fifty years, showing the lifespan and cost of maintenance. If we require experience figures, we may in the meantime almost starve, or suffer from the effects of increased nuclear waste radiation and pollutants thrown off in the burning of coal and oil. We now recognize that we have no idea of the longterm cost of nuclear installations. We can only guess at the cost of dismantling and entombing nuclear plants when they become inoperative.

Critics of OTEC are individuals without practical experience in the field. If in the early stages of nuclear power development scientists had been compelled to answer as many questions as are raised about OTEC, not a single nuclear installation would exist today. Lloyd Trimble, of Lockheed, notes that the first large plant will be costly, and adds: "If we have the guts to build a second anyway, we will build thousands." As with an automobile, the prototype is expensive.

There is no justification for arguing about which type of OTEC plant should be built. If fresh water is needed then, of course, the open cycle is necessary. An open-cycle plant can be used to produce fresh water only when the temperature is as low as 9° C or 10° C. Both closed and open systems have advantages. The site temperature differential of offshore currents, and whether or not fresh water is needed, as well as other conditions, can determine which type to build. Dr. Clarence Zener at Carnegie-Mellon and Dr. Hugh Sephton of the University of California are researching a "foam cycle." A column of foam pushes an upper column of water through a water wheel instead of a steam turbine. This method holds great promise, but it should not stop us from forging ahead with the other known methods. It is also possible to combine the open and closed cycles. Undoubtedly there will be more alternatives. That is the nature of scientific advancement.

A Hep Senator

United States Senator Spark Matsunaga [Democrat] of Hawaii wrote this past September [1978] to [former] Secretary of Energy James Schlesinger:

I have recently learned that the Department of Energy is presently planning on not requesting adequate funds to begin design and construction on a 10 to 20 megawatts ocean thermal energy conversion pilot plant during fiscal year 1980. It was my earlier understanding on the basis of hearings held before the Senate Subcommittee on Research and Development this past spring that DOE intended to propose funding for the start of the pilot plant during fiscal year 1980.

As you know, great strides have been made in OTEC technology during the past few years; workable solutions have now been proposed for all of the earlier defined problems, and needed hardware tests and preliminary design steps have been successfully completed on all major components.

Recently, the State of Hawaii and several firms began work on a mini-OTEC plant . . . financed largely with state and private funds.

Cost estimates of OTEC-generated power are most encouraging, and I fully expect this source to become economical and make a significant contribution toward meeting the energy needs of the nation in the near future.

Wonderland, USA

It is incredible that among the top levels of the Department of Energy there is no one with long-term experience in

OTEC and very few with an interest in the field. Officials without OTEC experience are evaluating Department of Energy and industry reports and making decisions. The field has been over-researched during its sixty years of hardware history. To some it looks as if the top echelon of the Department of Energy is stalling progress in this most promising solution to our energy crisis. When the Department asked TRW's Dr. Robert Douglas at a Los Angeles hearing if he could document the statement that his firm could build OTEC competitive with nuclear plants he replied: "Of course. I can send you a truckload of studies and documents."

With the exception of the automobile, government subsidies have been the sole mechanism enabling large-scale technology to develop consumer hardware. Unfortunately, our Department of Energy's present procedure is in contradiction to this. It will not pay the cost to implement any new technology and is limiting its assistance to advice and for specialized research and development. Regrettably, the government is likely to make decisions on the basis of political considerations rather than on technical feasibility. OTEC plants need single contractors able to build and integrate all components and to follow through.

Gamesmanship

Both [former] Secretary Schlesinger and President Carter showed their nuclear bias in the National Energy Plan of 1977. Both have had more experience with atomic energy [and] still hold to the dream of yesteryear that it will solve our problems. The government has lured utility companies to go nuclear by a complicated series of tax breaks and preferences. Accordingly, some electric companies felt almost compelled to make new investments in this area.

The Department of Energy should be accountable to the people. We pay the bills. It's time the President, Congress, and the Department of Energy know how we feel. The present course can very well lead to a breakdown in our economy. Food, transportation and energy distribution systems, waste disposal in our cities, and banking will be cracking, breaking down. Does anyone believe anymore that the poor nations that are getting poorer will ever be able to pay back their development loans?

Awake, DOE!

We can avert world depression and economic chaos if we immediately shift our priorities. Giving tax credits to those who use solar panels to heat swimming pools and incentives to oil drillers to investigate new sources of oil is like sending boxes of bandaids to an area devastated by a killer earthquake.

As I testified this past December [1978] before the Department of Energy:

This is the third time I have had the pleasure of testifying before you on solar energy. We do not have to wait for a depression. Hundreds of millions of people are on the edge of starvation or close to freezing. Their governments are nearing ruin and will pull with them large American banks committed to unrepayable loans. Yet with brave resolve and relatively small initial investment, we can turn that series of events around and make our nation the most beneficial ever to have graced this earth.

To say the least, it is disillusioning to find so little understanding within the federal bureaucracy. They prefer to pretend that all can proceed as usual, that with a little effort here and there, and giving incentives to the oil and gas folks, we will muddle through.

Working With Mother Earth

Now that I am in my ninth decade, I may be forgiven for reminiscing. Back in my native Norway, Mother Earth whispered—and sometimes thundered—to me from as far back as I can remember. My ancestors who came to these shores a thousand years ago found the native Americans to be kindred spirits. They intermarried, and they communicated with each other and Mother Earth. You would be amazed at the number of engineers, physicists, biochemists, and laborers who enjoy the privilege now. Today our earth is speaking more loudly than ever before, telling us of energy systems ready to go.

Understandably, four billion people would never decide on a single energy system. Multiple approaches are to be taken for granted. Let us proceed with centralized as well as decentralized earth-bound solar systems.

E. F. Schumaker, a vigorous proponent of decentralization, wrote in *Small Is Beautiful:*

For his different purposes man needs many different structures, both small ones and large ones, some exclusive and some comprehensive. Yet, people find it most difficult to keep two seemingly opposite necessities of truth in their minds at the same time. They always tend to clamor for a final solution, as if in actual life there could ever be a final solution other than death.

Yes, we need both centralized and decentralized energy systems to satisfy the needs of humanity. Let us continue with photovoltaics, capturing the wind, converting land and sea biomass into methane, harnessing tides, courting waves, and developing hydrogen technology.

People as Boss

Let's remind the Department of Energy that it is working for us. This is not the time for ponderous new studies and reports that will be subsequently evaluated by persons without experience in the field. Nor is there time to fritter away our human assets by concentrating all our efforts on the growing problems with oil and nuclear development.

Rather than listen to those who plop out with the popcorn wisdom that "solar energy won't make it in this century," we should look to the engineers who have worked with solar hardware and know what it can do. As a nation we are ready for a crash program in energy machines. We need not be afraid of crash programs. They are built into our temperament, and they may be most efficient. They brought us victories in World War II and in landing on the moon, and they are particularly applicable to OTEC.

TIDAL POWER IN THE BAY OF FUNDY[6]

Since the oceans formed, their tides have alternately
ebbed and flowed at shorelines in their diurnal wanderings.
Centuries ago, ingenious Europeans harnessed the energy of
tides to turn water wheels, converting the energy of moving
water into the spinning motions of useful machines. Today
several nations are examining the potential of tidal waters to
spin turbines and generate electricity, following the lead of
France and the Soviet Union, where practical tidal power
conversion machinery was built in the 1960s. Tides, as it turns
out, contain about 4×10^{12} watts of total power, an amount
that is by coincidence within an order of magnitude of the
world's present rate of electrical consumption.

The world's largest tides occur in the Bay of Fundy, which
separates New Brunswick and Nova Scotia in eastern Canada.
The head of the Bay, an extended ocean inlet about 200 miles
long and 40 miles wide, sees an average tidal rise and fall of
about 37 feet, and a phenomenal rise of 56 feet over the low-
tide mark has been recorded there. Each cycle of the Fundy
tide dissipates close to 5×10^8 kilowatt-hours of energy, a
quantity nearly equal in magnitude to the consumption of the
entire Canadian electrical network.

A plant harnessing some of this power could add to the
practical experience already generated by the French and
Soviet projects and might encourage other nations with
appropriate coastlines to extract power from the tides.
Cook Inlet in Alaska, the coastline of the English Channel
and the Irish Sea, the Sea of Okhotsk off the coast of the Soviet Union, the coast of the Yellow Sea off Korea, and the
inlets of the Kimberley Coast in Australia offer high tides and
convenient sites for constructing tidal power conversion machinery.

[6] Reprint of article by George F. D. Duff, professor of mathematics, University of
Toronto. *Technology Review.* 81:34–42. N. '78. Copyright © 1978 by the Alumni Association of the Massachusetts Institute of Technology. Reprinted by permission.

Why the Oceans Have Tides

The tides are moving masses of water, and by the laws of Newtonian physics must be subject to the influence of imbalanced forces. The tides are caused by spatial variation across the earth of the gravitational forces of the moon and sun. The lunar attraction pulls the oceans on the near side of the earth toward itself most strongly, creating a tidal bulge or wave in the oceans. The moon also pulls the comparatively rigid earth away from the oceans on the far side, thus raising a double bulge of ocean waters that follows the moon as it orbits the earth. Because the earth rotates once every twenty-four solar hours, this double wave system is perceived on earth to have a period of twelve hours plus a correction of twenty-five minutes to account for the moon's simultaneous progress in its orbit.

The gravitational attraction between earth and moon is not the largest force; that of the sun is larger. Only when the difference between forces at the surface and center of earth is taken does the moon's effect exceed the sun's. The vertical component of these differential forces is 10^5 times smaller than the gravitational attraction of the earth for its oceans, so the horizontal component is actually the force that does the work.

Because the sun's gravitational effects are far weaker on earth than are the moon's, the solar tidal wave system, which has the same double-wave configuration as the lunar system, has lesser magnitude. The period of the solar tidal wave system is perceived to be exactly twelve hours on earth, because we keep solar time. Note that the lunar tidal bulges are approximately aligned with the earth-moon line, as are the solar tidal bulges with an earth-sun line.

The ebb and flow of the tides is a phenomenon perceived only by an observer on an ocean shore. The horizontal tide-raising force has the ultimate effect of oscillating the oceans slightly—causing a rise on the order of magnitude of one foot at sea—thereby transferring energy to the affected masses of water. As the tidal waves, many hundreds of miles long, advance, the uplifted waters climb onto beaches and enter inlets, raising the observed sea level. Depending on the configuration of these inlets, the amplitude of the tidal wave may be increased many times. (The term *tidal wave* should not be confused with the tsunami seismic phenomenon sometimes incorrectly called a tidal wave.) As the earth continues to rotate, the areas flooded by high tides reemerge, and the water drains back to its lowest level as the tide ebbs. The motion of the flowing and ebbing tides represents the vast levels of energy already cited, which dissipates over periods of several days.

At new moon the moon and sun are aligned on the side of the earth nearest the sun; at full moon they are also in alignment, but the moon faces the side of the earth farthest from the sun. During both these lunar phases the lunar and solar tidal wave systems reinforce each other, producing high amplitude "spring" tides. But a week or so later from either phase, during half moon, the two tidal systems are 90° apart and interfere, causing lesser "neap" tides.

The Ebb and Flow of Tidal Technology

Currents generated by the ocean tides maintain speeds as high as seven to eight knots for several hours as they enter and leave some inlets, providing such remarkable displays of power as those in the Bay of Fundy. Here, because the tide is funneled between two land masses, such an incoming tidal wave can form an impressive wall-like buildup of water called a tidal bore. Tidal power has several major attractions:

☐ Tidal power is constantly being renewed but is not conservable because it disappears if not utilized within hours.

☐ Tides are an indigenous resource that if exploited would conserve dwindling fossil fuel supplies and would reduce problems of foreign exchange inherent in the importation of oil.

☐ Tidal power plants have extended lifetimes on the order of one hundred years, with extremely low and stable future operating costs.

☐ Tidal power is nonpolluting and has minimal environmental side effects.

Methods of extracting useful mechanical motion from rapid movements of water are centuries old. From the twelfth century until 1956, tidal energy turned water wheels in a mill on the Deben River at Woodstock in Suffolk, England. The mill is now an historic monument.

In more recent years, tidal energy has been successfully harnessed in several nations to generate electricity. In these nations and elsewhere tidal power potential has been studied in great detail.

Since 1966 the modern 240-megawatt-capacity tidal power plant on the Rance estuary in Brittany, France, has supplied electricity reliably to the French electric power grid. When spun by rushing tide water the turbines in the Rance installation activate electrical generators. Powered by electricity from the grid, the same turbines also pump ocean water into the Rance estuary during intervals near high and low tide. This pumping augments overall energy recovery by producing higher "heads" of water and increasing the period of productive gravity-induced flow through the turbines during the following part of the tidal cycle. Twenty-four of these beautifully designed 10-megawatt-capacity turbine pump-generators are installed in bulb-like casings, which are mounted in passageways in a half-mile-long dike spanning the estuary.

In 1967, Soviet engineers built a two-turbine tidal test plant at Kislaya Guba, an inlet of the White Sea, using turbines similar to those in the Rance installation. Their innovative method of installing the turbines was extremely successful. After preparing the site, the engineers installed the turbines in a concrete caisson, which was floated to the site and then sunk into position at low tide. This procedure of floating mounted turbines into position is now favored by designers in some other nations, including Canada, that contemplate building their own tidal power installations.

In the 1930s Passamaquoddy Bay, on the Atlantic Coast joining Canada and the United States, was the subject of preliminary tidal power works that were never completed; some dikes were built and abandoned. A recent Department of Energy study estimates fairly high costs compared to projected electrical output for any future installation there, because these tides are lower than in more promising sites. However, the geography of the islands and channels around Passamaquoddy Bay may be advantageous for storing tidal water in two huge basins. Such a double-basin system would enhance the extraction of power by extending the period of usefully ebbing tidal water. More may yet be heard of Passamaquoddy.

While Great Britain will almost certainly build some sort of system to extract power from ocean movements, for the moment it is not clear whether the system will be based on tidal power or wave power. A stiff competition between the two technologies for government interest and support exists in British energy circles, with strong and weak arguments for both sides. Tidal power certainly has great potential: calculations indicate that several-thousand-megawatt blocks of power could be obtained from a tidal power barrage built across the turning point in the Severn estuary, between England and Wales. But proponents of wave-power conversion point out that their systems can be much smaller than tidal power schemes and do not require the effort and expense necessary to dam entire bays and estuaries. The tidal power supporters point out that the efficiency of wave power installations is unpredictable, suffering when wave patterns are irregular and the weather is unfavorable.

An Optimistic Reassessment of Fundy Tidal Power

A joint Federal/Provincial Tidal Power Review Board (TPRB) concluded . . . [in 1977] that electricity produced by harnessing Fundy tidal power can now compete economically with electricity produced by other means, and would in fact displace the production of electricity by fossil-fueled generating plants. The review suggested that the integration of tidal power into the planned generation system of the Canadian Maritime provinces is both technically and economically feasible.

The TPRB's positive set of conclusions is the latest and most definitive in a series of joint Federal/Provincial government assessments of the potential for Fundy tidal power conversion. A 1969 study had recommended three sites along the Bay of Fundy as most promising for tidal powerplant development. The first site was at Economy Point in Minas Basin, Nova Scotia, near the world's highest semidiurnal tides. Tides at the other two sites—one at Cumberland Basin and the other at Shepody Bay near the border between New Brunswick and Nova Scotia—are of slightly smaller magnitude.

There was little incentive to act on the 1969 recommendations until the OPEC oil embargo struck in 1973, markedly improving the economic prospects of tidal power, and leading to new studies that culminated in the 1977 TPRB reassessment. Already the governments of Canada, Nova Scotia, and New Brunswick have planned a $30 million detailed design study for a tidal power plant in Cumberland Basin. It is expected that this design study will be the first major task of the newly formed Maritime Energy Corporation, which will oversee the development of the electrical system in Nova Scotia, New Brunswick, and Prince Edward Island.

The 1977 reassessment, while positive overall, was unusually complex and pointed out several areas needing further study. It included detailed discussions of technical problems related to harnessing the tides at the Cumberland Basin as well as calculations of the smooth integration of tide-generated electricity into the electrical system of the three provinces. The research topics included the following:

☐ determining the effect of a tidal power conversion plant on the magnitude of the tides in the Bay of Fundy;

☐ estimating the cost of the project and its probable economic impact;

☐ designing and building the installation;

☐ timing and absorbing the entry and exit of large blocks of electrical power to and from the Maritime provincial system; and

☐ forecasting and assessing environmental effects that could result from the installation of the tidal power conversion system.

The Tidal Resonance Problem

The Bay of Fundy has an overall resonant frequency, similar to the acoustic resonance of an organ pipe. This resonance is considered to be essential to the production of Fundy's very high tidal amplitudes. Just as changing the dimensions of an organ pipe "detunes" it so that it no longer produces high amplitude vibrations, building a tidal barrier would shorten the length of the Bay and diminish its natural resonance. If this upset were to seriously detune the Bay and reduce tidal amplitude significantly, the very existence of a tidal power conversion plant might paradoxically reduce its own effectiveness.

Both the 1969 and 1977 TPRB studies investigated the likelihood of such detuning effects. Numerical modeling techniques were used primarily to describe the rise and topography of the floor of the Bay, and the convergence of its shorelines, which together act to increase tidal wave amplitude greatly.

The amplitude of a tidal wave in deep ocean is rather modest—on the order of one foot. But as such a wave enters and advances up the Bay, it is increased by the rising ocean bottom and funneled by the converging shorelines. In the upper reaches the wave is steepened by the effect of the rising bottom, and then progresses up the river as a tidal bore, visually distinct from the relatively quiescent waters already there. The length of the Bay and velocity of the tidal wave are such that the rapid flow and subsequent ebb of the tide are complete just prior to the beginning of the next tidal flow.

Should the length of the Bay be changed so that an incoming tidal flow were faced with an ebbing tide partway up the Bay, its velocity would be reduced, and so would its rise in amplitude. A tidal wave following directly on the heels of a departing ebb tide would have the best chance of maintaining a high amplitude and producing a large tide at the head of the Bay. Resonant high tides will be most likely to occur if the natural period is about twelve hours, twenty-five minutes.

The TPRB considered that a tidal barrier constructed at a point well below the head of the bay will shorten its natural period. The decrease in period will throw off the timing, or resonance, of the ingoing and outgoing tides, leading to reduced tidal amplitude, and affecting tidal power prospects. For example, it was calculated that barriers at the giant sites near the tip of Cape Chignecto where Fundy divides would reduce the amplitudes there by one third, rendering these sites completely unsuitable for tidal power conversion. The TPRB recommended that the sites of tidal barriers should be located fairly close to the head of a bay to minimize the likelihood of such decreases.

An early calculation of the natural period of the Bay of Fundy itself produced a value of about nine hours—much smaller than actual field observations. This low value suggested that the observed resonance involved a larger sea area, which was subsequently found to include both the Bay of Fundy and also the Gulf of Maine, limited on the southwest by the shallows south of Cape Cod and on the northeast by the banks extending seaward from Nova Scotia. The shallow fishing ground of Georges Bank occludes much of the perceived sea boundary so that nearly all the tidal wave energy enters through the narrow Fundian Channel, a submarine valley 150 fathoms deep.

By comparing amplitudes of the principal lunar, principal solar, and lunar ecliptic tidal harmonics within the Bay of Fundy/Gulf of Maine system with corresponding amplitudes in the outer North Atlantic Ocean, C. J. R. Garrett showed that the Fundy/Maine system behaved as if its natural period was approximately 13 hours—about one hour longer than the optimal 12-hour period. This finding gave the first evidence of the possibility that a tidal barrier in the longer Minas Basin arm of the Bay might *increase* the resonance and actually *raise* tidal amplitudes by reducing the Bay's natural period slightly.

Detailed calculations by D. Greenberg have shown that a tidal barrier at Economy Point in Minas Basin can be expected to increase tidal amplitudes slightly, but at other points in the system rather than at the barrier site itself. This large site nonetheless would be the most economical of the possible sites because of its very high tidal amplitudes. Two potential sites in Cobequid Bay are probably below the resonant length so that barrier construction at either location would diminish the tidal amplitude up to 6 or 7 percent locally, but would not cause any appreciable increase elsewhere.

The Economics of Fundy Tidal Power Conversion

The TPRB estimated a capital cost of $3.6 billion (in 1976 Canadian dollars) for the installation of a 3,800-megawatt capacity tidal power conversion plant at Economy Point in Minas Basin—about the same capacity as four large nuclear powerplants. Such a plant would yield at-site power at a unit cost of 18 mills per kilowatt-hour. However, 3,800 megawatts is more power than could be integrated into the Maritime Provinces' system in the foreseeable future; the capital cost burden would therefore also be unnecessarily large. For these reasons the TPRB deferred further consideration of this potent tidal power site and turned its attention instead to the two favorable sites in Chignecto Bay. For these sites, projected at-site costs for tidal electricity were 30 mills per kilowatt-hour in Shepody Bay (1,550 megawatts capacity) and 22 mills per kilowatt-hour at Cumberland Basin (1,085 megawatts capacity).

What would the Maritime provinces get for such huge investments of capital? Beyond the electrical power generated by the tidal powerplant, significant fuel savings would also be realized. For example, the Maritime provinces now use fuel-oil-fired boilers to power most of their electrical generators. The costs of fuel oil rose steeply over the last several years. A TPRB computer model calculated that a tidal power plant at

Cumberland Basin could save three million barrels of fuel oil annually—as well as 380,000 tons of coal and some nuclear fuel. Such savings over a period of years might well exceed the costs of constructing the tidal power plant. For a thirty-year period, the cost/benefit ratio for a Cumberland Basin plant (and also for the larger Economy Point site) was estimated at 1:1.2. In arriving at this estimate, the TPRB used a "real" annual interest rate of 5.5 percent. In their calculation this rate was adjusted to remove the effects of inflation, but included realistic risk and borrowing costs. On the basis of these cost/benefit projections the TPRB selected the Cumberland Basin site for a tidal power plant estimated to cost $1.2 billion, and the three concerned governments accepted the recommendation.

Barring unforeseen developments, a detailed design for the Cumberland Basin plant may be completed in three years. A final decision on construction would then be made. Construction could involve three methods:

☐ the Soviet method of floating in bulb turbines mounted in caissons, referred to as "construction in the wet" in the construction industry;

☐ building power houses and sluices behind dewatered cofferdams—"construction in the dry behind cofferdams";

☐ building the power houses and sluices underground in a convenient headland and later removing natural rock plugs to open the passageways for the water—"construction in the dry behind natural rock plugs."

About ten years would be required to complete the construction of the plant, so that it could be expected to be operational by 1990.

Design and Operation of the Proposed Cumberland Basin Tidal Plant

The design now proposed for the Cumberland Basin plant uses a "single effect" tidal barrier, and a single basin to moderate the flow of tidal water. Single effect generation involves filling the enclosed basin using additional pumping, only at the end of ebb tide. The water would enter through 24 large sluice gates as the tide comes in and would exit through 37 7.5-meter diameter turbines, each with a rated generating capacity of 31 megawatts. The filling and emptying operations would be timed to make the most electricity from each tidal cycle.

Fitting Tidal Power Into the Provincial Electrical System

The problem of scheduling or retiming the periodic blocks of electrical power from the Cumberland plant and of absorbing it into the Maritime energy system involves many complexities. On some days the timing of the block of tidal electricity would fit in neatly with the daily demand peak. But the moon advances fifty minutes in its orbit each day, so that a few days later the power would be produced at the wrong time. The size of the tidal power block would grow gradually from neap to spring tides—a period of one week—and then decline. The TPRB developed a computerized model to study the integration of the tidal blocks of power and found that over 80 percent of the Cumberland Basin output could be directly absorbed in the Maritime Province system. The remaining energy could be stored for later use in the Maritimes, or exported to the New England Power Pool.

Geophysical and Environmental Factors

Perhaps the most serious environmental risk of a tidal power plant at Cumberland Basin is in its effect on the erosion, transport, and deposition of sediment in the Bay of Fundy, and of the subsequent effects these new patterns might have on life forms living along the Bay.

The strong Fundy tidal currents entrain vast tonnages of silt and coarser sediment about the upper branches of the Bay and especially in Minas Basin. Much of this fine-grained material is eroded from cliffs along the shoreline by the gradually encroaching sea. Sea level is creeping upwards in this area at a rate of about one foot per century. Much smaller-scale river barrier construction in the Maritime provinces has already caused unforeseen river narrowing and sandbank formation, and with these examples at hand, the TPRB is pursuing the continued study and modeling of sedimentation processes in the Bay of Fundy.

A newly established environmental assessment panel is charged with determining the effects that such changes in tidal regime and sedimentation patterns will have on fish and mudbank flora and fauna. While extreme changes in sedimentation patterns are unlikely to be caused by a tidal plant at the Cumberland Basin, the actual extent to which these populations will be affected at specific localities cannot yet be . predicted with any certainty. Overall limits of change can be estimated more reliably, and are smallest for the Cumberland Basin site itself.

The Future of Tidal Power

The proposed 1,085-megawatt capacity Cumberland Basin tidal plant represents a step forward in scale for tidal power technology. In addition, double-basin schemes to enhance the storage and facilitate the retiming of tidal flow are still possible in both upper branches of the Bay of Fundy. Such schemes could give the Maritime provinces an option of choosing among tidal, fossil-fuel, and nuclear power that does not yet appear at this early planning stage.

The potential of tidal power conversion, while perhaps less than that of solar, wind, and hydropower resources, still could become a substantial natural supplement to the economic use of renewable energy. Very full development of tidal energy conversion plants in the Bay of Fundy could possibly harness some 10 to 15 percent of the available natural tidal energy. On a worldwide scale, many less concentrated sites could capture and use perhaps 3 percent of the planet's tidal power, if the economics continue favorably.

GEOTHERMAL GOES EAST[7]

Texas tool pusher J. O. Gurley squinted at the hazy sun, disappearing in clouding sky over Fort Monmouth, New Jersey. Then he glanced at the scene before him: semi-trailers from Texas, Oklahoma, and Mississippi, loaded with casing and drill pipe. Powerful portable generators stood all around the grounds. At the center, tethered like a diabolical carnival ride, a specially designed mobile drilling rig towered ninety feet above the hundred-odd dignitaries, workers, and reporters. Gurley, fresh from Alaska's North Slope, spoke impatiently.

"Git these people out of here and git that fixed." He pointed to a corner of the field where five of Houston's best oil-field welders were repairing a platform damaged in transit. "If the rain holds off we'll be down and out in five days. Hell, hot water should be a piece of cake."

Drilling for hot water? Thirty miles from Times Square? Yes, Gurley and his crew are seeking a new source of geothermal energy, the ubiquitous but hard-to-get warmth of Mother Earth herself.

Geothermal energy has long been regarded as the exclusive preserve of the geologically dynamic Far West and a few foreign countries. Say "geothermal" and most people think of The Geysers area of northern California, or isolated hot springs in Nevada, Utah, and Idaho, or steaming sites in Iceland and Italy. But within the last few years, geologists have discovered potentially vast sources of geothermal energy in the United States, right in the area that needs it most: the industrial, populous eastern United States.

And there's more warming news. While Gurley and his crew were drilling in New Jersey, another crew was poking deep into the earth at a test area near Alvin, Texas, about twenty miles south of Houston. These two sites are virtually

[7] Reprint of article by Peter Britton, science writer. *Popular Science.* 241:66–9. F. '79. Reprinted from *Popular Science* with permission. © 1979 Times Mirror Magazines, Inc.

at opposite ends of a broad system of coastal plains extending down the Atlantic seaboard and westward along the Gulf. The Department of Energy (DOE) is providing most of the funds for the tests to find out if these zones of underground heat can yield usable energy.

Today, the US geothermal power output is about 520 megawatts—enough to heat, light, and cool a city of two million people. But if all goes according to Hoyle and not according to Murphy, one expert says the geothermal energy from these new eastern and Gulf Coast sources could amount to the equivalent of four million barrels of oil a day by the year 2000, or about 20 percent of our present oil imports. But, despite J. O. Gurley's opinion, getting it will be no piece of cake.

Atlantic Coast Geothermal

In the eastern target area, from Fort Monmouth to Savannah, Georgia, fifty shallow (1,000-foot-deep) test holes are scheduled to be drilled. The potential geothermal energy to be tapped here is different from that now used in the West in one important respect: It is relatively low temperature. In the West, steam from the earth and high-temperature water under pressure are used primarily to generate electricity. If useful geothermal heat is found along the East Coast, it will be too cool for electrical generation—but ideal for a number of low-temperature applications. In terms of the fossil fuel saved, it makes little difference, however.

These primordial heat sources originated some 200 to 400 million years ago when massive molten granitic intrusions called plutons squeezed up through other rock layers. Some protruded and formed the Appalachians, Poconos, and other mountains. Some remained submerged and gradually cooled. But other plutons were rich in radioactive uranium, potassium, and thorium, never completely cooling. In fact, the slow decay of these radioactive minerals assures, according to the theory, a steady and reliable source of heat. The heat is conducted to the top of the plutons and normally dissipates. But under the Atlantic coastal plain there is about a mile-thick blanket of shales, clays, and sandstones that acts as insulation, trapping the heat.

Geologists believe that there are major aquifer systems (porous sandstone formations that hold water) at the bottom of the pile of sediments covering the plutons. These would collect the trapped heat in water up to 225 degrees F. The DOE's idea is to drill into the formation and recover the water. Its heat would be removed and used in a variety of ways: residential and commercial space heating, animal-shelter heating, food processing, tobacco and pulp processing, greenhouse heating, crop drying, fish farming, for example. The water would then be pumped back into the ground.

The present test borings are being made to measure heat availability and conductivity and to make heat-contour maps. Measurements are taken after each hole has cooled for a month or so. The aim: to find the most promising locations for deep drilling. At present, the prime candidates appear to be the areas along the southern New Jersey coast, the Delmarva Peninsula, and around Norfolk, Virginia; Wilmington, North Carolina; Charleston, South Carolina; and Savannah, Georgia. Other information will come from gravity data, air surveys of magnetic fields, and other geologic studies. If the test holes show promise, a 4,000–7,000-foot hole will be drilled into an aquifer and, with luck, a pluton.

At Fort Monmouth I talked with Dr. Dave Lombard, the assistant regional manager for DOE's eastern geothermal-energy programs. He was on hand to watch the operations directed by DOE's drilling program manager, Glen Stafford, and carried out by Gruy Federal, an engineering firm in Houston.

It's possible that within a few years people right here in central New Jersey and elsewhere in the East could be using geothermal energy for heating and cooling their homes, apartment buildings, and businesses [Dr. Lombard told me]. If we find what we hope to find, we just may be able to deliver a reliable, competitively priced energy source, one that has been overlooked for years.

The United States is not the leader in the field of low-temperature geothermal energy. Similar geologic structures are yielding hot water for heating apartments and greenhouses in Hungary and France, among other places. Dr. Al Stone, director of advanced-research programs at Johns Hopkins University's Applied Physics Laboratory, has been conducting an extensive study on geothermal energy for the DOE and recently toured several European facilities. Among his observations:

☐ In Budapest 6,000 homes use geothermal domestic hot-water systems installed over the past thirty to forty years. The goal for Hungary is space heating of an additional 100,000 to 200,000 dwelling units.

☐ Thirty acres of greenhouses in Szentes, Hungary, use hot water from the earth to raise two vegetable crops a year.

☐ In France, some 17,500 apartments (mostly in the Paris Basin) now have geothermal space heating and another 4,600 earth-heated units are under construction. The target for 1985 is 500,000 dwellings.

☐ Some 4,000 apartments in Creil, France, save 4,800 tons of oil a year by using geothermal energy with titanium-plate heat exchangers and heat pumps.

Within a year, scientists should know a great deal more about the geothermal potential of the Atlantic coastal plain—whether, in fact, there's enough there to bother with and if it's technically and economically attractive.

Gulf Coast Geothermal

The Texas end of the line, however, is a bit further along and perhaps of greater overall importance. This geologically different area promises a different form of geothermal energy: hot saline water under tremendous pressure. And dissolved in it are large quantities of methane—natural gas.

For decades, oilmen working along the Gulf Coast have cursed these reservoirs of fiercely hot, pressurized salty water. When a drill bit was tapped into one, the water would flash into steam at the wellhead and in some cases blowouts would occur. These inadvertent strikes were downright dangerous and a waste of time and money. The oilmen learned how to drill safely into the formations for pockets of free methane. But the water-bearing zones were cased off and forgotten. Now these nuisances are viewed as a potential triple source of energy.

The idea is conceptually simple, if operationally difficult. First, extract the methane from the hot (up to 300 degrees F.) brine in a high-pressure separator, and dispatch it in ordinary pipelines. The remaining fluid, still very hot and under high pressure (up to 16,000 psi [pounds per square inch]), would go through a hydraulic turbine to produce electricity from the kinetic energy. The fluid would emerge from the hydraulic turbine and go through a heat exchanger to extract the heat. This could be a closed-loop system with a low-boiling point fluid (such as isobutane) that would be heated by the brine and run through a turbine to produce electricity. The brine would then be injected back into the earth. It might even be recycled to pick up more heat and gas.

There are several alternatives for the heat energy as well. Instead of producing electricity on the spot, the brine could be sent twenty or thirty miles by pipeline to an industry for direct use as process heat. Or the hot fluid could be used for space heating, cooling via absorption-type air conditioners, or for agricultural processes.

The idea that geopressured methane could be successfully recovered from these formations has been under examination, at least theoretically, for some time. But it came a large step closer to reality when a depleted and shut-in gas well in southern Louisiana was reopened last year. The geopressured zone was penetrated and fluid was produced at the rate of 10,000 barrels a day, yielding methane at up to eighty cubic feet per barrel. The success of this test has made the idea look very attractive. The area has been punctured by literally hundreds of thousands of holes. The plugged and abandoned ones are worthless now. But there are dry holes that are still unplugged and old wells that are just beginning to run dry, all potential sources of methane-bearing, energy-rich saline.

University of Texas geologists, working under a DOE contract, identified the Alvin, Texas, site as a promising area for a $3 million test well. Its official name is the General Crude Oil Company/Department of Energy Pleasant Bayou Well Number One. General Crude, the owner and driller, will drill and case the well to 16,500 feet, run a complete series of electronic logs, and go through a two-year-long production test. The prime objectives: to see if they can produce methane and hot water economically, and if so, for how long. To be economically viable the Pleasant Bayou Well Number One would have to produce 40,000 barrels of brine yielding 1.6 million cubic feet of natural gas per day.

According to Dr. Myron Dorfman at the University of Texas, three cubic miles of reservoir may be required to make the undertaking worthwhile. With luck, they might hit a 10-billion-barrel reservoir at 15,700 psi and 300 degrees F. with 400 billion cubic feet or more of methane in solution—the energy equivalent of 72 million barrels of oil.

Cost, Corrosion, Collapse

Why hasn't this been done before? The nation didn't need the energy until recently, and gas used to be only 16 to 20 cents per thousand cubic feet. Now it costs $1.42 and, as Dr. Lombard says, "We can afford to think about it." But there are other reasons why potential geothermal sources lay dormant, reasons of both nature's and man's doing. The extraordinarily hot environment wreaks havoc on equipment. The abundant minerals in the water can clog a two-inch pipe in less than forty-eight hours. Chemicals, caustics, and carbon dioxide also cause erosion.

The hot, salty water, up to three times saltier than sea water, can eat into even stainless steel. And the high pressure forces the brine through the pipe at twenty-two feet per second, another vicious eroding action, especially if grains of sand are carried by the brine stream. "Under some conditions," says Dr. Lombard, "such scouring could cause equipment failure, and a spectacular blowout."

And there is the worrisome problem of potential subsidence, or lowering of the ground level, even up to several feet in a few years. There is a great difference of opinion as to whether it will occur, but various schemes to detect any subsidence are being planned.

One method would fire radioactive pellets into the wall of the hole at one-hundred-foot intervals. Periodic testing with a Geiger counter-like instrument should indicate any vertical shifting. A second method would shoot transverses north-south and east-west to get the current ground level. Subsequent readings would show up even minor sinkage.

The University of Texas is working on a new device called a dual-liquid tilt meter. Engineers will lay long tubes on the surface, fanning out radially from the well to five hundred meters. Fluid levels at each end measure changes in elevations as slight as one tenth of a millimeter. As Dr. Dorfman says: "We'll know if a cockroach walks across the site."

Red Tape

There are other possible environmental problems, such as brine disposal, noise, and induced seismicity. And beyond nature's considerable reluctance in giving up her warmth is man's old habit of making things difficult for himself. Under the heading "institutional barriers" is a formidable list of complications: resource ownership, jurisdictional responsibilities, tax incentives, loan guarantees, depletion allowances, drilling permits, leasing of federal lands, zoning laws, impacts on local communities, and other arcane dealings.

For example, there is considerable debate at the moment about what the fluid—the geopressured brine—really is. Says DOE's director of geothermal energy, Rudolph Black: "Should this fluid be considered water or mineral? The laws are conflicting. Some states consider it water, and it therefore comes under surface-water rights and laws. But in other states it's a mineral and comes under the mineral laws. In yet other states it's unique unto itself with a whole set of laws that apply just to it. In many states it's unresolved."

On the positive side, however, the National Energy Act grants a special 10-percent depletion allowance for geopressured natural-gas wells drilled between October 1978 and December 1983, and investment tax-credits and other incentives for the development of all geothermal resources.

Although some gas experts doubt that very much of the geopressured methane can be recovered economically, University of Texas geologists feel that there could be as much as 250 trillion cubic feet of recoverable methane in the offshore and on-land Gulf Coast basin. Another group goes even higher, estimating that more than 800 trillion cubic feet might be available. The United States currently uses about 20 trillion cubic feet of natural gas each year.

If It Works . . .

Dr. Dorfman remains optimistic.

I emphasize that we have not yet proved the feasibility of the concept. But if it works—and there's a 75-percent chance of that within the parameters we have set—this triple energy resource could play a truly significant role in years to come. For there are more than two dozen geographically similar geopressured areas in the United States and around the world, in such places as the North Sea and Siberia. But it's not going to be an instant panacea.

It is comforting to think that there may be enormous energy reserves steaming away right beneath our feet. But, as is true of every alternate energy source, geothermal development is a study in potential, problems, and compromise.

EDITOR'S INTRODUCTION

"Conservation," Harvard's Energy Project concluded, "may well be the cheapest, safest, most productive energy alternative readily available in large amounts. By comparison, it is a quality energy source. It does not threaten to undermine the international monetary system, nor does it emit carbon dioxide into the atmosphere, nor does it generate problems comparable to nuclear waste. And contrary to the conventional wisdom, conservation can stimulate innovation, employment, and economic growth." (See "Needed: A Balanced Energy Program," by Robert Stobaugh and Daniel Yergin, in Section II, above.)

The two selections that make up this section explore that thesis, one that gained ground during the 1970s as more and more people began to view conservation not as a series of restraints but as a tool that could help free Americans from binding energy costs. Some experts have calculated that the amount of energy saved through conservation could equal more than all the energy supplies the nation imports.

Denis Hayes, now head of the US Solar Energy Research Institute, made that point graphically back in 1976, when he figured that $300 of ceiling insulation in the average house could save sixty barrels of oil in less than nine years. In this way, he concluded, "we are 'producing' heating oil at about $5 per barrel when we install ceiling insulation." (At the time, heating oil sold for $16 a barrel—more than $20 less than 1980s price.)

Conservation of this sort has been called "the good news about energy." In this section's first piece, John H. Gibbons and William U. Chandler of the University of Tennessee Environment Center report some of that news. They define three types of conservation: *curtailment* (doing without); *fuel*

198

switching (moving from oil, for example, to less costly coal); and adjustments such as adding insulation, for which they reserve the word *conservation.* Then Gibbons and Chandler suggest ways that conservation might be encouraged, holding the misconception that economic growth is inextricably linked to increased energy use up to the light of recent studies.

In "The Davis Experiment," the final article in this book, Associated Press reporter Jeanie Esajian takes a look at the nation's most conservation-minded city. Bike trails, building codes, and the determination of its residents to save energy have made Davis, California, a model for the rest of the nation. Conservation is practical right now, the Davis experiment has proved—and worth the effort.

CONSERVATION'S PROMISE[1]

Energy conservation became necessary during the oil embargo of 1973-74 because of shortages and sharply rising prices; consequently, most people have come to associate energy conservation almost exclusively with curtailment. In contrast to curtailment (heroic measures to reduce consumption quickly by whatever means are least costly), conservation is the economic response to total energy cost. It means changing technology and procedures to reduce the demand for energy (or for specific scarce fuels) without corresponding reductions in living standards. Conservation can be regarded as a means of enhancing our economic welfare, leaving society materially better off. Thus, conservation is an act of enlightened self-interest.

When the price of energy was much lower, uses that are now economically wasteful were economically rational. The manufacture and use of automobiles that provide only ten

[1] Reprint of article entitled "A National Energy Conservation Policy," by John H. Gibbons, director, and William U. Chandler, research associate, University of Tennessee Environment Center. *Current History.* 75:13–15+. Jl. '78. Copyright © 1978 by Current History, Inc. Reprinted by permission.

miles per gallon is a good example. Waste, in this sense, is not a vice in and of itself; it is an economic term. Some may object that unless the costs of energy are carefully and broadly defined, this definition of waste does not properly take into consideration the external costs of energy use (e.g., air pollution from automobiles). This objection is appropriate. The failure to internalize all the costs of energy production and consumption has created many of our most serious environmental problems and social inequities. Legislation like the Federal Clean Air and Water Acts, the Coal Surface Mining Control and Reclamation Act, and the Federal Mine Health and Safety Act has now brought us closer to internalizing the cost of energy use, as well as the use of most goods and services.

The implementation of a comprehensive National Energy Policy must incorporate total social costs of energy. Attention must also be paid to fuel switching (as a means of conserving scarce fuel or avoiding the use of particularly obnoxious fuel forms) which, with conservation and curtailment, completes the list of the means of reducing energy use.

The implementation of energy conservation policy has been hampered by the following misconceptions regarding the meaning and practice of conservation:

☐ Energy and the production of goods and services are intimately and inextricably linked; energy is a relatively fixed factor in the gross national product.

☐ Energy consumption and jobs are inextricably tied together. More energy consumption means more jobs and vice versa.

☐ Higher illumination levels generally help productivity; low illumination is injurious to the eyes.

☐ Reducing the growth of energy consumption implies the replacement of machines (e.g., bulldozers) with manual labor (e.g., men and women wielding picks and shovels to build interstate highways).

☐ Turning down the thermostat at night is counterproductive—the energy used in heating up the house in the morning more than offsets any savings.

Recent energy conservation research dispels these misconceptions and aids in understanding the potential of conservation over the next thirty years or so.

There are basically two types of responses to energy price increases and other indications of increasing scarcity: behavioral changes, and changes in existing as well as new energy-consuming equipment.

Behavioral changes include many of the actions we take under emergency conditions, like curtailing the use of automobiles, using energy-consuming equipment less, taking shorter showers, and so forth. These are short-term changes that can be made quickly and at little or no cost. Other changes, like lowering thermostat settings, are important and yield benefits in health and comfort (many buildings are not only overlighted but also overheated in winter and overcooled in summer).

Over the mid- and long-term, modifications in energy-consuming equipment can yield substantial savings in energy and money, because energy use in the United States is far from efficient. The inefficiency of a typical gas-heated home is an excellent example. In combustion as much as 25 percent of the energy in the gas goes uselessly up the stack. The pilot light can consume 10 percent of the gas used in a home. Losses through poorly insulated heating ducts can amount to 40 percent. Altogether, from 35 to 55 percent of the useful energy in the natural gas can be wasted in a typical gas furnace system. The situation is similar in homes heated by oil and electricity. In addition to the inefficiency of delivery of heat into the residence, there are two other major sources of loss. First, the thermal shell of the residence is often very transparent to heat and cold, and thus loses energy. Second, the "availability" of heat from these furnaces (typically more than 1,000° F.) is inherently vastly greater than required—thereby incurring further losses.

The possibility for conservation in industry can amaze even the proponents of conservation. A recent *Wall Street Journal* article reported the following vignettes of energy savings investments in industry:

☐ One company invested $30,000 in boiler controls: savings from this investment equal $60,000 per year, a 200 percent annual rate of return.

☐ A tire manufacturer spent $40,000 to insulate steam valves: savings amount to $80,000 per year, another 200 percent per year rate of return.

☐ General Electric invested $13,000 at one of its plants to shut down certain equipment automatically: annual savings are $51,000.

These are easy and obvious efforts; however, the number of conservation actions are apparently very large.

Incorporation of energy efficiency into new energy consuming products, like new houses, can yield very large benefits. In fact, careful design and construction can do more than conserve energy. Double-paned glass, thicker insulation, and other energy conserving investments in buildings may decrease the total investment in the structure, for example, by decreasing the size of the air conditioner or furnace that must be provided. Due to this offsetting effect, large improvements in energy efficiency can be obtained with small net changes in total investment.

A prudent consumer would measure the benefit of his expenditure on various energy-saving options and decide which would yield the highest value to him. Educational programs modeled after the highly successful Agriculture Extension Service to give the consumer helpful information about intelligent investments in energy-intensive products are being tested in ten states.

Industry can respond to increasing energy prices in a variety of ways; thus industrial response requires detailed examination. Aluminum production, for example, can be reduced in energy intensity by nearly 40 percent, while chemical production may be able to manage only a 20 percent reduction. Uranium enrichment, the process by which nuclear reactor fuel and weapons material for the military is made and which consumes 4 percent of all the electric power in the United States, can be reduced in energy intensity by 90 percent. These improvements can come only over a time span corresponding to the replacement of existing production plants, and as a response to increases in energy prices.

Another example that highlights the potential for conservation in new equipment and illustrates the problem of achieving such savings is what we may call "the paradox of the automobile." To increase gas mileage while retaining a certain amenity level, that is, interior space, safety, and other essential features, we must pay the price of adding improved design and high-technology features like microprocessors (minicomputers) for fuel combustion control and more expensive but lighter-weight materials, like aluminum, or other innovations. Thus, there is a trade-off between fuel and nonfuel costs. The dilemma consists of choosing the right point between the two extremes of fuel and nonfuel costs. The total cost of owning and operating an automobile (the sum of fuel and nonfuel costs) is remarkably insensitive to fuel economy over the broad range of performance from 15 miles to nearly 35 miles per gallon. In other words, it makes little economic difference to the buyer to choose a 35-mile-per-gallon car instead of a 15-mile-per-gallon car. But one can argue that it is in the national interest to choose the more efficient car and to build a more sophisticated domestic car. Though not in the purest sense a paradox, it is puzzling that a choice so profoundly important on the social level is of such little consequence on the individual level. Clearly, the market process does not enable consumers to discern the most efficient choice in terms of merged national and individual self-interest.

Are there many such market distortions? In energy, the answer unfortunately is yes. In the past, the government instituted many regulations and incentives that still provide major subsidies to energy production and use in the United States economy. These include natural-gas price controls, tax incentives, and loan subsidies for energy production. New federal regulations and incentives and federal oil policy can alter this situation. For example, Congress has mandated minimum automobile average gas mileage (under the Energy Policy and Conservation Act of 1975, 1985 cars must obtain an average of 27 miles per gallon). Tax rebates and low cost

loans can be provided for homeowners who insulate their
homes. New construction standards for improved energy efficiency are being promulgated. But perhaps more important,
subsidies, like the subsidies for the transport of foreign oil to
the United States, should be stripped away from general tax
revenue. Only when all the costs of energy production are included in the price of energy (that is, internalized) will price
signals indicate the need to induce high energy efficiency utilization.

Implications for Economic Growth, Employment, and Technology

One way to try to understand the impact energy conservation may have on the critical national issues of economic
growth, employment, and the future of technology is through
the "scenario." Scenario analysis has been widely used to as-
sess the impacts of certain policy choices on our energy future. One such effort constructed a series of scenarios of different energy use levels in the United States to the year 2010,
using a combination of econometric and engineering analysis
techniques. In one low-energy growth scenario, total energy
consumption in the United States is projected to be about 30
percent higher in 2010 than it is now. This glimpse of an
imaginary but plausible future assumes that:

☐ Real income doubles by 2010, a growth significantly lower
than obtained in 1950–1975.

☐ Real energy prices steadily increase and double by the
year 2010, corresponding to an average 2 percent increase
per year. State and federal policy strongly supports actions to
increase the efficiency of energy use.

☐ Population increases by about one third (Bureau of Census
Series II projections).

Results of the analysis indicate that an economically rational response to these conditions could result in the following possibilities:

☐ Energy use in buildings could decline at an annual rate of
6 percent, compared to an annual rate of increase of 3 per-

cent currently (in part due to tough building codes and appliance efficiency standards).

☐ Electricity could account for 30 to 50 percent of US energy consumption in 2010, as compared with 28 percent today, depending on policy actions and the relative prices of other forms of energy.

☐ Natural gas could supply 11 percent of total demand, compared with 24 percent now.

☐ Auto efficiency doubles, in part due to regulations.

☐ Energy consumed per unit output of industry in 2010 is 35 percent lower than in 1975.

In other words, energy utilization efficiencies are made possible by the intelligent response of advanced technology to high and rising energy prices.

Perhaps the single most important conclusion of this study is the statement that the economic well-being of the United States can be maintained while energy growth is diminished. In this projection, gross national product grows at a long-term average annual real rate of about 2 percent (higher in the near term and lower in the long term). This rate reflects the expected slowdown in the growth of the labor force, because of the decreasing birth rate, and the continued long-term shift in the economy to service-oriented activity. If GNP [Gross National Product] grows faster, then more energy will be required (e.g., an average 3 percent annual GNP growth will require about 35 percent more energy than an average 2 percent annual growth).

As for employment, the study indicated no major difference in the demand for workers between the lowest and the highest energy demand projection considered. Moreover, energy conservation does not imply the substitution of manual labor for energy, since 20 cents worth of energy will buy the equivalent of a day's work of one laborer at a wheelbarrow. It therefore does not seem likely that we will opt for a return to wheelbarrows, picks and shovels! Instead, highly technical skills, those of electronics technicians, skilled construction workers, engineers, and others who can provide ingenuity in place of the brute force consumption of energy and other resources, will be required.

Consideration of a national energy conservation policy begins with a definition of conservation, including three kinds of conservation: curtailment, fuel switching, and what we call conservation itself. Curtailment involves heroic measures to reduce energy consumption quickly by the cheapest means available. Fuel switching is a method of conserving specific fuels and can be an important complement of conservation. *Conservation* is the term we reserve for the policy of substituting new technology or different procedures for energy without reducing the amenities we enjoy. Conservation in this sense—an economic sense—is a means of leaving society better off than it would be without it, and is thus an act of enlightened self-interest.

Changing our energy-using behavior can be accomplished much more quickly than can the substitution of new and more efficient energy-consuming equipment. Lowering thermostat settings is a behavioral change that can yield large savings. With the exception of modifications or retrofits to existing property, the replacement of a more energy-efficient house or car must await the end of the useful life of the house or car. The rate of capital equipment turnover, therefore, controls to a large extent the rate at which conservation can be implemented. As a means of dealing with near-term shortages, policymakers can focus on behavioral changes or curtailments, but they must view the retrofitting of old equipment and the incorporation of energy efficiency into new energy-consuming goods as ultimate goals. There are many opportunities of both types in today's economy because the thermodynamic limits of efficiency are far from actual practice.

Ideally, the marketplace would determine the optimum investment in conservation. But the "paradox of the automobile" highlights the distortion that can occur in the price signals consumers see. Despite the profound national need for increased automobile efficiency, it makes little difference to the individual whether he buys a 15-mile-per-gallon car, or one that obtains 35 miles per gallon. Policy tools include stripping away subsidies (for example, subsidies on the importation of foreign crude oil) of energy consumption, providing incentives for conservation, and internalizing all the costs of energy consumption. [As of April 1980 the tax credits for the industry still exist.—Ed.]

Will Conservation Stifle the Economy?

Will the conservation of energy stifle the economy, constrain technological innovation, and reduce employment? Research indicates that there need be no net difference between a high energy-use future and a low one in the demand for workers. The economy can double its present size by the year 2010 while energy use in that year can be held to a level no higher than today's. Furthermore, it will be necessary to substitute ingenuity for brute force in the consumption of energy, which means that more highly skilled workers (and fewer workers performing menial tasks) will be required.

The realization of these positive aspects of energy conservation, however, is contingent upon action to convert our energy-consuming capital stocks soon enough to avoid the necessity of major changes in consumption patterns. Such changes may otherwise be in the offing a decade or more before the end of the century.

THE DAVIS EXPERIMENT[2]

In much of the country, energy conservation is something people may think or talk about. In this northern California university town of 35,000, it's become a way of life.

Citizens here use bicycles rather than cars, heat and cool with solar panels, recycle waste, grow vegetables in city-supplied plots, and make the insulated, energy-efficient home a

[2] Reprint of article entitled "In One Town, Energy Conservation Is a Way of Life," by Associated Press reporter Jeanie Esajian. *Gannett Westchester Newspapers.* p B 6. N. 27, '79. Reprinted by permission of The Associated Press.

point of pride. A strict city code reinforces volunteer efforts.

No wonder President Carter singled out Davis recently as a model community for an energy-pinched time, noting that the city has done "a tremendous job" in slashing energy use.

Davis claims an 18 percent drop per household in electricity consumption and a 27 percent drop in household gas consumption since 1973, when a series of unusual planning and building codes were enacted. About one fourth of the population travels on bikes, cutting gas consumption.

The bike habit started on the University of California campus and spread to the city. The "waste-not" philosophy also derived much of its impetus from the university community.

In 1972, with the help of the eighteen-year-old vote, three of four campus-backed candidates were elected to the five-member city council. They favored slow, planned development and preservation of the rural atmosphere. And they helped put through building ordinances and a housing allocation system centered on energy conservation.

The rules set building specifications that resulted in rigorously insulated homes and gave priority for construction to builders with the most energy-efficient house plans.

The code essentially requires more insulation for walls and ceilings, limits the number and size of windows, most of which must face north and south, and requires a certain portion of window space to be shaded [says Doran Maxwell, acting community development director].

The code allows for some flexibility. If a homeowner wants more windows, for example, he must trade off with more insulation—water-filled columns or containers to retain heat or cold and a concrete floor slab covered with tile or linoleum.

Hearings also are under way on new rulings which would require some outfitting to existing homes before they could be resold, says Maxwell.

Dennis Forsberg, a regional planner working on a study of the Davis ordinances, says the city's success is due both to the 1,300 energy-efficient housing units built since the codes were enacted and public awareness and information.

It was Davis' reputation as a conservation-minded city that attracted Maria Buckman and her family, after years of $100-a-month utility bills with their thermostat at 85 in summer and 70 in winter.

Now their bill averages $32 in the winter and $19 in the summer.

Average temperature range in Davis in July is 56–95; in January, 39–53.

"It's really very simple," she says of life in Davis. "It reminds me of how my grandparents lived. We take a few steps backward and I think we'll go forward."

The Buckman house, part of a large development of solar-powered homes, has solar collectors on the roof to heat water and uses "passive" techniques or design elements to reduce use of conventional energy by either keeping out unwanted sunlight or putting it to work.

Fifty-five-gallon oil drums in front of south-facing windows are exposed at night and retain coolness during the day after awnings are drawn to block the sun. Double-pane window glass and heavy-duty insulation keep cool air from leaking out. A vegetable garden provides the family with food and helps absorb heat.

While it's 113 degrees outside, it's 82 degrees inside their house with only natural air conditioning.

Houses in the development start in the low $60,000s and go up to $90,000, says developer Marshall Hunt, who with his wife, Virginia Thigpen, builds solar homes in Davis.

They have succeeded under the city's new building codes, but, on the other extreme, some builders "have had to leave town," Ms. Thigpen says.

The pair stress that solar collectors and other elaborate devices are less important than basics like good insulation and proper arrangement of windows. In their own "passive" solar home, which stays at 75 to 78 degrees in summer, their July electricity bill was $5.37. In winter, with rare use of an electric heater, their highest bill is $12, they say.

If all houses had those basic elements, it would cut down on energy consumption 50 percent [Hunt says]. Although it isn't as glamorous to think about things like insulation, weather stripping

or a more efficient refrigerator than solar collectors on the roof,
they go 75 percent of the way to solving the problem.

Solar homes cost no more than conventional ones, he says.

"You can save money on central heating and air conditioning units and you save with the state's solar tax credit," he
says. California's solar tax credit allows a homeowner to de-
duct 55 percent of the cost of the system from income taxes.

While many people move here and settle in new energy-
saving housing developments, others have gone to great
lengths to install solar devices in their existing homes. This is
called "retrofitting."

Terry Lyons heats the water for his home and his back
yard hot tub with a solar system installed in his 1969 duplex
by next-door neighbor Steve Byars for $2,200. Anybody with
"reasonable mechanical skills can build these things," Byars
says.

Although the collectors are more expensive than those
built into a new home, Byars thinks they're worth it. "We're
within half a year of payoff on our own solar system," he says,
"and we look forward to thirty more years in which it'll work
for us."

The Pacific Gas & Electric Company office in Davis has
joined the trend with a $40,000 system supplying all the hot
water and 70 percent of the interior heating for the 4,450
square-foot building in which twenty people work.

Company official Bill Hoppert says the system was developed for "testing and teaching" and is not expected to be cost
effective. For business, Hoppert says, the utility can't endorse
solar systems from a cost standpoint except for water heating,
"which is very close to being cost-effective."

If many Davis residents conserve gas and electricity at
home, even more conserve their share of gasoline by riding
bicycles. Always big on the University of California campus,
the trend widened in a town that's compact and flat.

With the fall influx of 17,500 students, Davis' bike census
reached 35,000, or one per resident. The city has 35 miles of
bike paths and lanes, and there are 21 miles on campus.

The spirit of conservation in Davis extends beyond utilities and transportation.

Many residents grow their own fruits and vegetables in community gardens. That indirectly saves on transport energy.

Recycling is also big in Davis, where the city mandates it for paper, cans, and glass. "It's a break-even thing," says Paul Hart, of the Davis Waste Removal Company. He figures it costs as much to truck the material to buyers as he gets paid for it. The service is subsidized by city garbage rates.

Still, for conservation-minded Davis, it all adds up, even where an operation isn't precisely cost-effective. The main point of much of the city's effort is to save energy, and here it's clearly showing the way.

BIBLIOGRAPHY

An asterisk (*) preceding a reference indicates that the article or part of it has been reprinted in this book.

BOOKS AND PAMPHLETS

Behrendt, Rees, ed. Symposium on Science, Technology and the Human Prospect. Pergamon Press. '79.

Bittlingmayer, George. The federal government's energy policies. 1978–1979 high school debate analysis; American Enterprise Institute for Public Policy Research. 1150 17th St. N.W. Washington, DC 20036.

*Brown, L. R. and others. Running on empty: the future of the automobile in an oil short world. Norton. '79.

Commoner, Barry. The politics of energy. Alfred A. Knopf. '79.

Congressional Quarterly, Inc. Energy policy. Congressional Quarterly Press. '79.

Gibney, Frank, ed. Energy: the fuel of life. Bantam/Britannica. '79.

Hayes, Denis. Energy: the case for conservation. (Worldwatch Paper 4) Unipub. '76.

*Stobaugh, Robert and Yergin, Daniel, eds. Energy future: report of the Energy Project at the Harvard Business School. Random House. '79.

Union of Concerned Scientists. What you should know about the hazards of nuclear power. The Union. 1208 Massachusetts Ave. Cambridge, MA 02138. [unpaged leaflet '79]

PERIODICALS

Atlas. 25:12+. S. '78. Atlas survey: energy—what options? ed. by Robert Bendiner.

Atlas. 26:27–9. O. '79. Future of solar power. Anil Agarwal.

*Bulletin of the Atomic Scientists. 35:13–17. Je. '79. Nuclear energy: what went wrong. C. L. Wilson.

Bulletin of the Atomic Scientists. 36:23–30. F. '80. Energy: a summary of the CONAES [Committee on Nuclear and Alternative Energy Systems] report. Harvey Brooks.

Bulletin of the Atomic Scientists. 36:55–9. F. '80. Conservation is here. Lee Schipper.

Business Week. p 55–6. Ja. 30, '78. Shale: the answer to domestic oil supply?

Business Week. p 27–8. Ap. 16, '79. Critical problems of nuclear power.

Business Week. p 68+. Ap. 23, '79. Gloom behind the natural-gas bubble.

Business Week. p 80–2. Jl. 16, '79. Quicker we get at synthetic fuels, the better we're going to be; interview by M. G. Sheldrick and John Love. C. C. Garvin.

Business Week. p 42+. Ag. 20, '79. Tough job of forging a Good Neighbor energy policy: North American Energy Common Market. S. W. Sanders.

Business Week. p 52–3. F. 25, '80. Watering down of energy proposals.

Center Magazine. 11:70–5. My. '78. Radioactive waste disposal—the key to a nuclear future; address, September 29, 1977. J. G. Speth.

Center Magazine. 11:32–45. S. '78. Soft energy path: with questions and answers. A. B. Lovins.

Changing Times. 32:6–9. Ap. '78. Solar heating your house—would it pay?

Commentary. 67:27–39. Je. '79. Harrisburg syndrome. Samuel McCracken.

Commentary. 68:61–7. N. '79. Solar energy: a false hope. Samuel McCracken.

Commonweal. 106:390–3. Jl. 6, '79. American auto, R.I.P. Frank Getlein.

Consumers Research Magazine. 63:27–8. F. '80. Solar energy [bibliography].

Current History. 74:97–138. Mr. '78. The world energy crisis; symposium.

Current History. 74:193–229. My. '78. America's energy resources: an overview; symposium.

°Current History. 75:1–38. Jl. '78. America's energy policy tomorrow; symposium.
  Reprinted in this volume: A national energy conservation policy. J. H. Gibbons and William Chandler. p 13–15+.

°Environment. 20:32–8. Mr. '78. Future for hydropower: small dams, non-dams. G. S. Erskine.

Environment. 20:11–15. Ap. '78. Economic comparison of three technologies: photovoltaics, nuclear power, co-generating engines. Robert Scott.

Environment. 20:17–20. O. '78. Power from the ocean winds. D. R. Inglis.

Environment. 20:4–5+. N. '78. Can high technology solve our energy problem? K. H. Hohenemser.

Esquire. 90:44–5+. D. 19, '78. Mexico's oil boom and what's in it for us. Christopher Buckley.

*Environment. 21:25–30+. Ja. '79. Why not methane? papers from Scientists' Institute for Public Information seminar.
Reprinted in this volume: Delivering methane. Eugene Luntey. p 33–6.

Environment. 21:6–15+. N. '79. Optimal solar strategy. S. J. Nadis.

Family Health. 11:6+. Jl. '79. Nader on nukes: the America syndrome. Ralph Nader.

Forbes. 123:91–2+. Ja. 8, '79. Energy's clouded future. James Flanigan.

Forbes. 124:27. Ag. 6, '79. Methanol age is dawning. James Flanigan.

Foreign Affairs. 57:836–71. Spring '79. After the second shock: pragmatic energy strategies. R. B. Stobaugh and D. H. Yergin.

Foreign Policy. 35:170–9. Summer '79. Oil is still too cheap. A. L. Madian.

Fortune. 98:50–2+. N. 20, '78. New fears surround the shift to coal. Tom Alexander.

Fortune. 99:62–4+. Je. 18, '79. Dealing with those windfall profits. Edward Meadows.

*Fortune. 100:110–11. Ag. 13, '79. Synthetic fuels can be economic now. W. M. Brown and Herman Kahn.

Futurist. 12:332–3. O. '78. When the oil will run out; views of Workshop on Alternative Energy Strategies.

*Gannett Westchester Newspapers. p B 6. N. 27, '79. In one town, energy conservation is a way of life. Jeanie Esajian.

Harper's Magazine. 259:16–18+. O. '79. The meltdown that didn't happen. Howard Morland.

*Humanist. 39:12–19. Jl. '79. Ocean thermal energy conversion. Bryn Beorse.

International Wildlife. 8:20–7. My. '78. Firewood: the poor man's burden. E. P. Eckholm.

Los Angeles Times. p V6. Ap. 29, '79. Looking for lessons beyond Three Mile Island: we need renewable energy. Barry Commoner.

Mechanix Illustrated. 75:32–3+. Mr. '79. 5-cent solar cell: Ovshinsky cell. C. A. Miller.

Mechanix Illustrated. 76:72+. Ja. '80. Sun and politics; the White House installation. C. A. Miller.

Mechanix Illustrated. 76:40–1+. F. '80. Ocean power hits Hawaii [Mini-OTEC system].

Mother Earth News. 58:94–5. Jl. '79. Informal directory of the wind energy industry. Mike Evans.

Nation. 228:417+. Ap. 21, '79. Kicking the petrol habit; gasohol—
a 100-proof solution. F. J. Cook.

*Nation. 228:521+. My. 12, '79. Nuclear power: the price is too
high. Richard Munson.

National Parks & Conservation Magazine. 52:16–21. Ap. '78. Endless energy: the long and short range of solar power. K. P.
Maize.

*National Parks & Conservation Magazine. 52:10–15. My. '78.
Harnessing the wind. Lee Stephenson.

National Parks & Conservation Magazine. 53:15–20. Je. '79. Coal:
saviour or demon? J. R. Boulding.

Nation's Business. 66:66–7. My. '78. Big ifs in coal's future. C. D.
Holmes.

Nation's Business. 66:78–80+. S. '78. Energy: searching for substitutes. Grover Heiman.

New Republic. 178:25+. F. 25, '78. Soft and hard energy. Nicholas
Wade.
    Same. Current 203:49–53. My. '78.

New Republic. 181:14–16. Jl. 21, '79. Synfuel madness. Morton
Kondracke.

New Republic. 182:15–17. F. 2, '80. Hostages of energy [U.S. consumption of petroleum]. Daniel Yergin.

New York. 12:64–9. Jl. 9, '79. Who's to blame for the high price of
oil. Anthony Sampson.

New York Review of Books. 26:14–17. My. 17, '79. Doing without
nuclear power. Charles Komanoff.

*New York Times. p A 12. Jl. 16, '79. Nation's energy chain: fuel
sources intertwine. A. J. Parisi.

New York Times. p 1+. O. 20, '79. Soviet reports major step toward a fusion plant. Theodore Shabad.

*New York Times. p 1+. D. 8, '79. Despite U.S. nudges, solar en-
ergy moves slowly. A. J. Parisi.

New York Times. p A 27. Ap. 17, '80. The Soviet-oil alarum. M. I.
Goldman.

*New York Times Magazine. p 32+. Je. 4, '78. The real meaning of
the energy crunch. D. H. Yergin.

Newsweek. 93:63–4. F. 26, '79. Power to save. M. Sheils and
others.

Newsweek. 94:23–6+. Jl. 16, '79. Why we must act now. Michael
Ruby and others.

Newsweek. 95:68. Ja. 28, '80. Studious nod to nuclear power [Na-
tional Academy of Sciences study].

°Oil and Gas Journal. 77:47–51. Mr. 26, '79. Coal's problems to keep pressure on oil and gas. Bob Tippee.

°Parade. p 4–5. F. 18, '79. Nuclear fusion: where to get energy when the oil wells run dry. Isaac Asimov.
Same with title: Nuclear fusion for energy: will it solve the oil crisis? Current. 212:39–43. My. '79.

°Popular Science. 214:66–9. F. '79. Geothermal goes East. Peter Britton.

Popular Science. 214:74–5. My. '79. Wave-tuned plates harness sea power. David Scott.

Popular Science. 216:30+. F. '80. Owner-built solar system for northern climes. P. M. Palmer.

Reader's Digest. 115:92–6. D. '79. Why the senseless rush to synfuels? David Stockman.

°Reader's Digest. 116:90–2. Ja. '80. Let's get the facts. James Nathan Miller.

Redbook. 154:50+. F. '80. Is nuclear power the only choice for the future? Judith Viorst.

Saturday Review. 6:24–6. Mr. 3, '79. Solar eclipse: our bungled energy policy. S. E. Ferrey.

Science. 199:634–43. F. 10, '78. Photovoltaic power systems; a tour through the alternatives. Henry Kelly.

Science. 199:756–60. F. 17, '78. Tar sands: a new fuels industry takes shape. T. M. Maugh, 2d.

Science. 199:879–82. F. 24, '78. Residential natural gas consumption: evidence that conservation efforts to date have failed. R. L. Lehman and H. E. Warren.

Science. 199:1041–8. Mr. 10, '78. Solar biomass energy: an overview of U.S. potential. C. C. Burwell.

Science. 200:142–52. Ap. 14, '78. U.S. energy demand; some low energy futures; report of the Demand and Conservation Panel of the Committee on Nuclear Power and Alternative Energy Systems.

Science. 203:233–9. Ja. 19, '79. Energy resources available to the United States, 1985–2000. E. T. Hayes.

Science. 203:252–3. Ja. 19, '79. Policy review boosts solar as a near-term energy option. L. J. Carter.

Science. 203:718–23. F. 23, '79. Risk with energy from conventional and non-conventional sources. Herbert Inhaber.

Science. 203:849–51. Mr. 2, '79. Energy choices for the next 15 years; a view from Europe. C. P. L.-Zaleski.

Science. 203:1214–20. Mr. 23, '79. Economic feasibility of solar water and space heating. R. H. Bezdek and others.

Science. 204:1069–72. Je. 8, '79. Petroleum exploration; discouragement about the Atlantic Outer Continental Shelf deepens. R. A. Kerr.

Science Digest. 85:77–81. Ja. '79. 21st century's energy source. Joseph Lippert.

Science News. 115:387–8. Je. 16, '79. Tapping heavy crude: a forgotten resource; first International Conference on the Future of Heavy Crude Oil and Tar Sands.

Science News. 116:45–6. Jl. 21, '79. A nuclear watershed. Janet Raloff.

Science News. 116:52–3. Jl. 21, '79. Reaching for the stars; fusion reactors. D. E. Thomsen.

Scientific American. 239:70–6. D. '78. Fuel-cell power plants. A. P. Fickett.

Scientific American. 241:72+. Jl. '79. Coal: prospects and problems; OTA report.

Scientific American. 240:38–47. Ja. '79. World coal production: Workshop on Alternative Energy Strategies study. E. D. Griffith and A. W. Clarke.

Sierra. 63:8–10. Jl. '78. How Sweden saves so much energy. James Keough.

Sierra. 64:42–4. Ja. '79. Biomass energy: the promise and the problems. William Lockeretz.

Sierra. 64:31–5+. Mr. '79. How much do you know about energy? Keith Kline.

Technology Review. 80:62–6+. Ag. '78. Can we save energy by taxing it? J. F. Boshier.

*Technology Review. 81:34–42. N. '78. Tidal power in the Bay of Fundy. G. F. D. Duff.

Technology Review. 81:58–74+. F. '79. Mining earth's heat: hot dry rock geothermal energy. R. G. Cummings and others.

*Technology Review. 81:68. Mr. '79. Coal: the ace-in-the-hole that isn't there. J. I. Mattill.

*Technology Review. 81:31–3. Ag. '79. The current politics of "synfuels." Philip Shabecoff.

Technology Review. 82:83–4. D. '79/Ja. '80. Solar conversion: U.S.-Soviet see-saw. J. I. Mattill.

Time. 112:91–2. O. 16, '78. Oil: what's left out there; Rand Corporation study.

*Time. 113:66–8. Ap. 16, '79. Use less, pay more; Carter's new energy plan.

Time. 113:70–4+. My. 7, '79. Inside the big oil game.

Time. 113:72–3+. Je. 11, '79. Energy: fuels of the future.

Time. 114:27. Jl. 2, '79. Possibility, not a novelty; Carter's solar energy program.

Time. 114:23. Jl. 9, '79. How to counter OPEC; Time essay. Marshall Loeb.

UNESCO Courier. 31:4–33. Je. '78. Energy for tomorrow's world; symposium.

*USA Today. 107:48–50. S. '78. Energy from biomass—coming full circle. H. L. Breckenridge.

*USA Today. 108:17–19. S. '79. The case against oil price decontrol. H. M. Metzenbaum.

USA Today. 108:14–15. Ja. '80. Curing a national paralysis [need for petroleum conservation]. M. K. Udall.

U.S. News & World Report. 85:33. N. 27, '78. Long-term outlook: steady climb in oil prices; interview. Ulf Lantzke.

U.S. News & World Report. 86:68. My. 7, '79. We've been asked: What can be done with atomic waste?

U.S. News & World Report. 86:43–4. My. 21, '79. Windfall profits tax on oil? interviews. S. E. Eizenstat; J. F. Bennett.

U.S. News & World Report. 86:26+. Je. 4, '79. What's holding up the switch to coal; interview. J. D. Rockefeller, 4th.

U.S. News & World Report. 81:18. Jl. 9, '79. Synthetic fuel: how helpful, how soon?

U.S. News & World Report. 87:38. Ag. 13, '79. How two nations make oil from coal.

U.S. News & World Report. 87:21–2+. Jl. 30, '79. How Carter plan would change life in the U.S.

U.S. News & World Report. 88:8. Ja. 28, '80. How to stretch U.S. energy to 2000 [study by National Academy of Sciences].

Vital Speeches of the Day. 45:261–4. F. 15, '79. Slowdown: the only answer to the energy problem? address, November 20, 1978. Ulf Lantzke.

Vital Speeches of the Day. 45:418–20. My. 1, '79. National energy policy: decontrol; address, April 5, 1979. Jimmy Carter.

Vital Speeches of the Day. 45:437–9. My. 1, '79. Zig-zag down and out: solar energy policies of the Department of Energy; address, March 22, 1979. R. B. Peterson.

*Vital Speeches of the Day. 45:642–5. Ag. 15, '79. Energy problems: the erosion of confidence; address, July 15, 1979. Jimmy Carter.

*Vital Speeches of the Day. 45:645–9. Ag. 15, '79. An energy secure America: $140 billion from windfall profit tax; address, July 16, 1979. Jimmy Carter.

*Vital Speeches of the Day. 45:764–8. O. 1, '79. The systems ap-
    proach to energy; address, August 15, 1979. A. B. Trowbridge.
Vital Speeches of the Day. 46:205–9. Ja. 15, '80. Energy for you
    and me, but what about the kids? [address, November 8, 1979]
    F. W. Lewis.
Washington Monthly. 10:42–6+. Jl. '78. Gas rationing: it worked
    before and it can work again. N. R. Burnett.
Washington Monthly. 10:50–8. Ja. '79. How Israel got the bomb;
    question of missing uranium from Nuclear Materials and
    Equipment Corporation. J. J. Fialka.
Washington Monthly. 11:50–7. My. '79. Missing link: why we
    can't stand up to OPEC. Fred Mann.